Ronald Antonio Rodriguez Moncada
Carlos Alexander Mendoza Jacomino

Transportation systems and urban planning

Ronald Antonio Rodriguez Moncada
Carlos Alexander Mendoza Jacomino

Transportation systems and urban planning

Efficient mobility and sustainable urban planning

ScienciaScripts

Imprint
Any brand names and product names mentioned in this book are subject to trademark, brand or patent protection and are trademarks or registered trademarks of their respective holders. The use of brand names, product names, common names, trade names, product descriptions etc. even without a particular marking in this work is in no way to be construed to mean that such names may be regarded as unrestricted in respect of trademark and brand protection legislation and could thus be used by anyone.

Cover image: www.ingimage.com

This book is a translation from the original published under ISBN 978-613-9-46908-6.

Publisher:
Sciencia Scripts
is a trademark of
Dodo Books Indian Ocean Ltd. and OmniScriptum S.R.L publishing group

120 High Road, East Finchley, London, N2 9ED, United Kingdom
Str. Armeneasca 28/1, office 1, Chisinau MD-2012, Republic of Moldova, Europe
Managing Directors: Ieva Konstantinova, Victoria Ursu
info@omniscriptum.com

Printed at: see last page
ISBN: 978-620-8-41365-1

Foreword

In a world where urbanization is advancing at an unprecedented pace, cities have not only consolidated their position as the nerve centers of economic and social activity, but also as scenarios of complex and urgent challenges. Problems such as road congestion, inequalities in access to mobility services, and the effects of climate change require comprehensive and innovative solutions that enable sustainable urban development. Transport Systems and Urban Planning: Innovation, Sustainability and Development, with the subtitle "Transforming the Cities of the Future: Efficient Mobility and Sustainable Urbanism", is a work that responds to this need, providing an in-depth analysis and visionary proposals to address these challenges. As Gehl (2010) points out, "cities must be spaces designed for people, where human interaction and access to services are the main focus of their planning" (p. 45). In this work, the author adopts this people-centered approach, offering practical tools and revolutionary ideas that not only solve technical problems, but also promote social and environmental well-being. From the implementation of electric transportation systems to the creation of inclusive public spaces, each proposal in this book reflects a commitment to sustainability and equity. The reader will find in these pages a comprehensive analysis that combines theory and practice. For example, Litman (2021) stresses that urban planning must adopt a multidimensional approach that considers sustainable mobility as a key element in reducing social and economic inequalities in contemporary cities. This principle guides the solutions put forward in the text, where not only technological aspects, such as intelligent transportation systems, but also social challenges, such as equitable access to mobility, are addressed. It is important to mention that the transformation of our cities cannot be the sole responsibility of urban planners or engineers. As Newman and Kenworthy (2015) emphasize, "urban sustainability is a collective effort that depends on active collaboration between governments, local communities, and private actors" (p. 87). This book echoes this idea, offering readers not only a deeper understanding of urban issues, but also an invitation to actively participate in building a more sustainable future. For professionals in the sector, this book represents an invaluable tool for understanding global trends and the most effective strategies in urban planning and transportation. For academics and students, it is a rich source of knowledge and references that frames current challenges in a practical and applicable context. For citizens interested in becoming agents of change, this book will provide them with clear and motivating ideas on how their daily decisions can positively influence their communities. As Jane Jacobs (1961) says, "successful urban development is that which respects the diversity and complexity of cities" (p. 25). This book not only respects that diversity, but also celebrates it, offering a framework for thinking about innovative solutions that will allow our cities to flourish in the 21st century. May these pages not only inform you, but also inspire you to act, because urban planning is not just a technical discipline: it is a collective act of imagination and creation that shapes the quality of life for millions of people.

Table of Contents

Introduction.......... 4

Chapter 1: Introduction to Transportation Systems 6

Chapter 2: Design of Sustainable Transportation Systems 20

Chapter 3: Urban and Land Use Planning 36

Chapter 4: Smart Technologies and Sustainability in Urban Transportation..... 52

Chapter 5: The City of the Future: Innovation and Sustainability in Urban Planning 71

References 87

Introduction

Urban planning and transportation systems are the fundamental pillars on which modern cities are built. Throughout history, cities have evolved from small settlements to megacities facing unprecedented challenges, such as population growth, climate change, traffic congestion, pollution and the need to ensure accessibility for all. In this scenario, civil engineering becomes a crucial discipline, and the book *"Transportation Systems and Urban Planning: Innovation, Sustainability and Development"* has been designed to guide current and future civil engineers in creating more efficient, inclusive and resilient cities.

This book covers everything from the fundamental principles of mobility and urban planning to the technological and sustainable advances that have revolutionized both fields. Readers will explore key concepts such as electric public transport, bicycle lanes, and innovations in shared and autonomous transportation, while gaining a comprehensive view of sustainable urban planning. In addition, the design of public spaces, the integration of green infrastructure, and urban densification strategies that improve the quality of life in the face of population growth are highlighted. Each chapter is enriched with contemporary examples and case studies, offering a practical and realistic perspective.

This book is not just a theoretical compilation, but a practical and visionary guide that prepares civil engineers to address current challenges and anticipate future ones. Beyond the technical aspects, it explores how urban planning and transportation systems interact synergistically to design functional and sustainable cities. It also analyzes disruptive trends and technologies, such as artificial intelligence in traffic management and micro-mobility solutions, which redefine accessibility and mobility standards. It also emphasizes sustainability and climate change, highlighting practices such as the use of renewable energies and sustainable transportation methods, essential to reduce the carbon footprint. In addition, it addresses urban policies and regulations, providing a comprehensive understanding of the legal frameworks, fiscal incentives and investment strategies necessary for the success of each project. Finally, the book offers an in-depth look at future projections, anticipating advances in autonomous transportation and new demands for urban resilience.

Aimed primarily at civil engineering students and professionals in the field, this book is also valuable for architects, urban planners, public managers and anyone interested in understanding the functioning and structure of modern cities. It is an essential tool that transcends technical knowledge, connecting engineering to an inclusive, resilient, and sustainable vision of the future. Through this book, readers will embark on a learning journey that will enable them to design and build cities that are more humane, ecological, and ready for the challenges of the 21st century.

Chapter 1: Introduction to Transportation Systems

At the heart of every dynamic and prosperous city, transportation systems are the fabric that connects its inhabitants, facilitating daily life, promoting the economy and reinforcing social cohesion (Garcia, 2023). Urban mobility, in all its forms, represents much more than the simple task of moving from one place to another; it is a transformative force that drives development and accessibility. In this first chapter, we will delve into the exciting world of transportation systems, exploring their historical evolution, their impact on modern life, and the various ways in which they shape our cities.

The study of transportation systems goes beyond infrastructure; it also encompasses how individuals and communities interact with their urban environment. From public mass transit to private and shared modes of transportation, each component of a mobility system has its own role in creating an efficient and accessible urban ecosystem (Lopez, 2022). This chapter explores the foundations of that structure, looking at the trends that have redefined mobility in recent decades and analyzing how transportation directly influences the socioeconomic development of cities.

Why is it critical to understand transportation systems? For civil engineers, understanding transportation systems is not just a technical requirement; it is a strategic necessity. By delving into the design and planning of these systems, engineers have the power to influence the daily lives of millions of people, optimizing their time, increasing their quality of life, and creating a more accessible and equitable environment (Martinez, 2021). This chapter provides a comprehensive overview of the different types of transportation systems, their particularities and how they are integrated into the urban context to provide mobility solutions adapted to the changing needs of cities.

Transportation systems as a tool for social change In addition to facilitating movement, transportation systems are powerful tools for social inclusion. A well-designed transportation system allows people from all backgrounds to access educational, employment, and recreational opportunities, promoting a more just and diverse city (Perez, 2020). In this section, we will explore how accessibility and inclusion are central axes in the creation of an effective transportation system, and how these principles help strengthen the social fabric.

Evolving mobility: current trends and future challenges

With the advancement of technology and the growing concern for the environment, transportation systems are constantly evolving. Today, more than ever, cities are looking for sustainable and efficient alternatives, such as electric public transport and infrastructure for non-motorized modes, that reduce the carbon footprint and promote the health and well-being of citizens (Gómez, 2023). This chapter also discusses the challenges facing transportation systems in a changing world: from managing vehicle congestion and pollution to integrating new technologies and the impact of shared mobility systems.

A glimpse into the future of urban mobility

Finally, we will delve into the future projections and changes that lie ahead in the field of transportation systems. We explore how artificial intelligence, autonomous vehicles, and mobility-as-a-service (MaaS) models are redesigning our expectations and possibilities for mobility in the cities of the future (Rodriguez, 2024). As urban needs evolve, civil engineers are called to anticipate these transformations and design transportation systems that not only meet today's demands, but are also prepared for tomorrow's challenges.

Welcome to an in-depth exploration of transportation systems, where technical knowledge is combined with social and environmental vision. In this first chapter, you will discover how urban mobility becomes an essential tool for the progress of cities and how, as future civil engineers, you can be part of this transcendental transformation in urban life. Join us in this initial immersion and discover the power of a well-designed transportation system!

1.1 Importance and Evolution of Transportation Systems

Transportation systems constitute one of the essential elements for the integral development of cities. Their evolution and adaptation have been fundamental to improve not only mobility, but also the quality of life, the economy and social development in urban areas (Litman, 2021). In the current context, transport systems are considered critical infrastructures, as they enable interaction between people, goods and services, strengthening the urban fabric and promoting social cohesion (Banister, 2018). The planning and management of these systems are, therefore, priorities for civil engineers, who must face the challenge of designing safe, accessible, sustainable and resilient systems.

Historically, the evolution of transportation systems has been marked by the needs of each era. During the industrial revolution, cities experienced rapid growth and drove the expansion of rail and tramway networks, which facilitated the movement of large populations to areas of work and commerce (Sclar, 2019). These developments enabled not only the connection between urban and rural areas, but also the growth of cities as centers of economic and social activity, thus creating the embryo of modern metropolises (Knowles et al., 2020). This expansion of urban infrastructure brought with it a new era of mobility that is still influencing urban design patterns today.

The second half of the 20th century marked another milestone in the evolution of transportation systems, with the rise of the automobile and the construction of highways in many cities around the world. This vehicular approach responded to the demand of a society that prioritized speed and autonomy in travel (Gossling, 2021). However, over the years, reliance on automobiles began to generate significant problems, such as traffic congestion, increased air pollution, and unequal access to transportation, especially in areas farther from urban centers (Mokhtarian & Chen, 2020). These negative effects drove interest in sustainable alternatives that prioritize public transport, active mobility (cycling and walking) and integrated multimodal transport infrastructures.

In the last two decades, attention has turned to sustainability and the reduction of the environmental impact of urban transportation. In this sense, electric public transport systems and the implementation of bicycle lanes have been promoted to encourage non-motorized mobility, initiatives that reduce both greenhouse gas emissions and noise levels in cities (International Energy Agency [IEA], 2022). The shift towards sustainable transport has been accelerated by international agreements on climate change, such as the Paris Agreement, in which countries committed to reduce global emissions to mitigate the effects of global warming (United Nations, 2015).

In addition, the development of technology has played a crucial role in this transition. Today, the integration of technology in transportation, such as traffic management systems, the use of mobility-as-a-service (MaaS) applications and autonomous vehicles, is transforming urban mobility, facilitating access and optimizing the use of available resources (Docherty et al., 2018). These advances allow traffic and resources to be managed more efficiently, offering greater convenience and accessibility for users, while reducing the environmental footprint of cities.

No less important is the role of transportation systems in social equity. A well-planned and accessible transportation system enables people from all walks of life to access employment opportunities, education, and essential services (Pucher & Buehler, 2019). In many cities around the world, the lack of transportation options in certain areas limits economic and social development opportunities for their inhabitants. Thus, transportation becomes a tool for social justice and inclusive development, and its planning and design must consider these aspects for the benefit of the entire population.

In conclusion, the evolution of transportation systems has been shaped by the social, technological and environmental needs of each era. Today, these systems face the urgent need to adapt to an era that demands sustainability, inclusiveness, and efficiency (Fernandez, 2023). For civil engineers, understanding this evolution and the current challenges is essential to design transportation infrastructures that not only respond to mobility demands, but also improve the quality of life of urban dwellers and contribute to the sustainable development of cities.

1.2 Types of Transportation Systems: Public, Private and Shared

In modern cities, a diversity of transportation options is essential to ensure efficient, accessible and sustainable mobility. Transportation systems can be divided into three main categories: public, private, and shared. Each of these types offers different benefits and challenges, and their appropriate combination can optimize the functionality of cities, reduce traffic, and mitigate environmental impact (Litman, 2021). In this section, we will explore the characteristics and role of each type of transport, as well as their influence on urban planning and social development.

Public Transportation

Public transport is an essential component of urban mobility and a key tool for sustainable development. Its ability to move large numbers of people via buses, trains and streetcars makes it an effective solution to reduce congestion and pollutant gas emissions (Banister, 2018). In particular, electric public transport systems have gained popularity in recent decades, as they contribute to decreasing carbon dioxide emissions and improving urban air quality (International Energy Agency [IEA], 2022).

In addition to environmental benefits, public transportation promotes social equity by providing an accessible means of transportation for people of all socioeconomic classes. As Pucher and Buehler (2019) point out, an efficient public transportation system enables low-income people to access employment, education, and health opportunities, regardless of their location within the city. Therefore, civil engineers must consider these aspects when designing and planning public transportation systems, ensuring that the service reaches all urban areas and that it is safe, efficient, and accessible to all.

Private Transportation

Private transportation, represented mainly by automobiles, has historically been the preferred method of mobility in many cities, especially in those that prioritize road infrastructure over mass transit (Gossling, 2021). The expansion of highways and the rise of the automobile during the 20th century marked a turning point in the configuration of many cities, consolidating a model of mobility centered on the private vehicle (Docherty et al., 2018). However, this dependence on the automobile has generated multiple problems, such as vehicular congestion, air pollution, and a growing need for parking space.

In recent decades, increasing traffic and its negative effects have driven the search for sustainable alternatives to private transport, and many cities are implementing policies to reduce car dependency. These policies include congestion charging, parking restrictions, and promotion of alternative modes of transportation, such as cycling and public transport (Litman, 2021). Even so, private transport remains a necessary option for certain contexts, especially in suburban and rural areas where public transport infrastructure is limited or non-existent. Thus, the challenge for civil engineers lies in balancing private mobility needs with sustainability and urban accessibility objectives.

Shared Transportation

Ridesharing has emerged as an innovative alternative that combines the advantages of public and private transportation. This type of transport includes services such as carsharing, ride-sharing and micro-mobility, such as shared bicycles and electric scooters. Ridesharing is a flexible solution that reduces the number of vehicles on the streets, reduces parking demand, and contributes to cleaner and more efficient urban mobility (Cohen & Shaheen, 2018).

According to Shaheen and Chan (2016), ridesharing offers significant environmental and economic benefits by reducing reliance on the private automobile and optimizing resource use. In addition, many studies point out that micromobility, especially in densely populated urban areas, can reduce greenhouse gas emissions and promote more active and healthy living (Banister, 2018). However, the implementation of ridesharing requires proper urban planning that includes charging and parking stations for bicycles and scooters, as well as policies to regulate their use and prevent problems of congestion on sidewalks.

Multimodal Mobility: Integration of Transportation Types

An emerging trend in urban transportation planning is multimodal mobility, which integrates various types of transportation (public, private and shared) to create a more cohesive urban mobility system that is adaptable to the needs of citizens. This integration allows people to combine different modes of transportation in a single trip, optimizing time and costs. An effective multimodal system offers flexible transportation options that can be adapted to each trip, improving the overall efficiency of urban mobility (Docherty et al., 2018).

The concept of mobility as a service (MaaS) is a modern manifestation of multimodal mobility, where digital platforms allow users to access multiple transportation options from a single application (Gossling, 2021). This not only facilitates the use of shared and public transportation, but also improves accessibility for people who would otherwise rely on private transportation. For civil engineers, designing infrastructure that encourages multimodal mobility involves planning interconnected transport stations and integrated payment systems that optimize the user experience and encourage a more balanced use of different types of transport.

The variety of transport types -public, private and shared- constitutes an integral system that, when properly managed and planned, can positively transform urban life. Each type of transport has its own role, and their effective combination can maximize the efficiency and sustainability of urban mobility (Ramirez, 2022). For civil engineers, understanding the particularities and advantages of each system is fundamental to create inclusive, equitable and sustainable cities. This comprehensive approach is key to addressing the challenges of congestion, pollution, and accessibility in the cities of the future.

1.3 Socioeconomic Impact of Urban Mobility

Urban mobility has a profound impact on the socioeconomic development of cities. Access to efficient and accessible means of transportation directly influences economic productivity, social cohesion and equal opportunities in urban areas. In essence, a well-planned mobility system acts as an enabler of economic and social growth, as it enables people to access jobs, education, health services, and leisure opportunities (Litman, 2021). This section explores the role of urban mobility in the economy, its relationship to quality of life, and how it can reduce or widen social inequalities.

Urban Mobility and Economic Growth

Efficient transportation systems foster economic growth by facilitating the movement of people and goods, enabling a steady flow of commercial and productive activity. Urban mobility optimizes travel times, which increases productivity and generates significant economic savings for businesses and workers (Banister, 2018). In urban economies, where the cost of time lost in traffic represents billions of dollars annually, improving transportation infrastructure is essential to maximize the competitiveness of cities (Rodrigue, 2020). According to a World Bank study (2019), access to efficient transportation can increase employment rates by connecting workers to areas that offer greater job opportunities, especially in metropolitan regions.

In addition, transportation facilitates the connection between markets, increasing accessibility to goods and services, which boosts demand and strengthens economic growth. In this sense, civil engineers play a crucial role in designing transportation systems that optimize urban logistics and improve the competitiveness of cities. Transportation investments not only boost the local economy, but can also attract foreign investment and foster tourism, key elements for the development of a strong urban economy (Gossling, 2021).

Social Equity and Accessibility

Urban mobility not only affects the economy, but also has a significant impact on social equity. Accessibility to affordable and accessible urban transport is essential to ensure equal opportunities in the city. An inclusive transportation system enables low-income people to access essential services, such as education and health, and reduces dependence on geographic location to take advantage of job opportunities (Pucher & Buehler, 2019). In particular, public transportation plays a fundamental role in social inclusion, as it enables mobility for those who do not have access to a private vehicle or cannot afford other transportation alternatives.

The lack of adequate transportation infrastructure in certain areas, especially in low-income neighborhoods and on the outskirts of cities, can create significant barriers to socioeconomic development. According to the United Nations (2021), people who lack access to adequate transportation face higher rates of unemployment and poverty, and are less likely to improve their living conditions. Therefore, equity in access to mobility should be a key component in urban transportation planning, and civil engineers should consider these aspects when designing infrastructure to ensure accessibility for all.

Public Health and Quality of Life

The relationship between urban mobility and public health is another area of significant impact. The design of a sustainable and healthy transportation system can reduce the negative effects of traffic, such as air pollution, noise, and traffic accidents, factors that directly affect the quality of life of urban dwellers (Mokhtarian & Chen, 2020). The World Health Organization (2021) has noted that exposure to high levels of air pollution and sedentary lifestyles, both influenced by car dependence, are associated with an increase in respiratory and cardiovascular diseases.

Promoting active transportation alternatives, such as walking and cycling, not only reduces greenhouse gas emissions, but also promotes a healthier lifestyle, which contributes to improving the health of the urban population (Cohen & Shaheen, 2018). For civil engineers, integrating active mobility options into urban design, such as bikeways and pedestrian zones, is key to promoting healthy mobility and reducing the negative impacts of automobile dependence.

Impact on the Environment and Sustainability

Urban mobility also has a direct impact on the environmental sustainability of cities. Urban transport systems are responsible for a significant proportion of greenhouse gas emissions, especially in those cities that rely heavily on private transport (IEA, 2022). The implementation of sustainable transport systems, such as electric public transport networks and infrastructure for active mobility, can significantly reduce emissions and contribute to sustainable development goals.

In cities that prioritize sustainable mobility, transportation becomes a tool for climate change mitigation and environmental protection. This translates into urban policies that promote the electrification of public transport, the reduction of motorized vehicles in the city center, and the expansion of green infrastructure (Banister, 2018). For civil engineers, the creation of sustainable transportation systems is a commitment to the environment and an opportunity to design cities that respond to today's environmental challenges.

The socioeconomic impact of urban mobility is vast, ranging from economic and social equity to public health and environmental sustainability. Each of these aspects highlights the importance of a well-planned and managed transportation system. For civil engineers, understanding these impacts is critical to designing transportation infrastructure that not only improves urban mobility, but also promotes economic development, social inclusion, health, and environmental sustainability. In an era of urbanization and climate change, urban transportation systems have the potential to transform cities and improve the quality of life of their inhabitants.

1.4 Current Trends in Mobility and Urbanism

Urban mobility and urban planning are undergoing a significant transformation, driven by technological changes, growing environmental awareness and the need for sustainability in cities. Current trends reflect a restructuring of traditional forms of travel and urban design, integrating innovative approaches that promote efficient mobility, accessibility, and environmental friendliness (Litman, 2021). This section examines the main trends in mobility and urbanism, from the rise of smart transportation and micro-mobility to the design of sustainable and resilient cities, highlighting their impact on the future of urban planning and the role of civil engineers in these changes.

Micromobility and Active Mobility

Micromobility, which includes the use of bicycles, electric scooters and other light vehicles, has become a growing trend in modern cities, as it offers a fast and environmentally friendly solution for short distances (Banister, 2018). This type of mobility not only reduces traffic on the streets, but also contributes to the reduction of emissions and promotes a more active and healthy lifestyle. Micromobility infrastructures, such as bike lanes and charging stations for electric scooters, are increasingly in demand in cities that prioritize sustainability and accessibility.

Active transportation, such as walking and cycling, has also gained popularity as part of a healthy mobility approach. According to the World Health Organization (2021), promoting active transport in urban areas can reduce sedentary-related diseases and improve air quality. This type of mobility requires safe and adequate infrastructure, such as wide sidewalks, pedestrian zones, and dedicated bicycle lanes, elements that civil engineers should consider when planning accessible and safe urban spaces.

Development of Compact Cities and Efficient Use of Land

The trend toward compact cities has strengthened in recent years, with a focus on the development of dense, multifunctional urban areas that reduce dependence on the automobile and optimize land use (Rodrigue, 2020). This urban model promotes the design of neighborhoods where residents can access basic services, such as schools, supermarkets, parks and offices, within a proximity radius that facilitates walking and cycling. The idea of the "15-minute city" is an example of this trend, proposing that residents can satisfy most of their daily needs just 15 minutes from home (Moreno et al., 2021).

The development of compact cities not only reduces the carbon footprint, but also improves quality of life by fostering social interaction and creating more connected communities. However, designing these areas requires careful urban planning that avoids congestion and ensures the availability of adequate infrastructure, from green spaces to public transportation networks. For civil engineers, this approach involves integrated and coordinated planning that considers efficient land use, controlled densification and the preservation of public spaces.

Sustainability and Urban Resilience

Sustainability and resilience are now key concepts in urban planning, driven by the need to mitigate the effects of climate change and prepare cities for extreme events such as floods, droughts and heat waves (United Nations, 2015). Sustainable cities seek to reduce their ecological footprint by using renewable energy, recycling waste, promoting sustainable mobility and designing resilient infrastructure.

To achieve effective resilience, many cities are investing in green infrastructure, ranging from parks and gardens to green roofs and sustainable urban drainage systems (SUDS). These elements not only help absorb rainwater and reduce flood risk, but also improve air quality, create habitats for biodiversity, and provide recreational spaces for citizens (Litman, 2021). Civil engineers play a key role in the implementation of these infrastructures, ensuring that cities not only respond to current challenges, but are also prepared to face the future effects of climate change.

Current trends in mobility and urbanism are marked by a search for sustainability, innovation and equity. These transformations require a civil engineering approach that adapts to the new demands of modern cities and integrates advanced technology, sustainable infrastructure, and inclusive urban design (Hernández, 2023). For civil engineers, understanding and applying these trends is essential to building cities that are not only efficient and accessible, but also promote the long-term health, well-being, and resilience of their inhabitants. By anticipating these trends, engineers can lead the development of urban environments that respond to the challenges of the future, laying a solid foundation for smarter, healthier and more sustainable cities.

1.5 Future Projections and Challenges in Transportation and Urban Planning

As cities continue to grow and face complex challenges such as climate change, congestion, pollution and the need for greater social equity, urban planning and transportation systems must adapt to meet the demands of a changing world. Civil engineers are uniquely positioned to influence the future of mobility and urban development, leading innovative and sustainable solutions that improve the quality of life in cities. This section explores future projections in transportation and urbanism, as well as the challenges facing cities and civil engineering in this context of accelerated change.

Urban Growth and Mobility Demands

Demographic projections suggest that by 2050, about 70% of the world's population will live in urban areas, increasing mobility demands and straining the capacity of existing infrastructure (United Nations, 2018). This urban growth creates an urgent need to plan transportation systems capable of moving large numbers of people efficiently and sustainably. Civil engineers must design infrastructures that not only meet current demand, but also have the capacity to adapt to future growth.

The growth of cities requires a mobility approach that moves away from reliance on the private automobile and promotes public transport options and active mobility. According to the World Health Organization (2021), high-capacity public transport systems and walking and cycling infrastructure are essential to reduce congestion and mitigate the environmental and social impacts of urban growth. In addition, the design of compact, mixed-use cities, where essential services are located close to residences, is a key strategy to reduce mobility demand and promote sustainability.

Autonomous Mobility and Connectivity

Automation and connectivity are transforming urban transportation, and in the future, autonomous vehicles are expected to play an important role in city mobility. Autonomous vehicles have the potential to reduce congestion and improve road safety, as they can communicate with each other and adjust their routes in real time to optimize traffic (Docherty et al., 2018). However, the introduction of autonomous vehicles also presents challenges, such as adapting road infrastructures and regulating their use in urban spaces.

Connectivity is also manifested in the use of mobility-as-a-service (MaaS) platforms, which allow users to access a variety of transportation options from a single application (Cohen & Shaheen, 2018). This multimodal and flexible approach improves accessibility and enables more efficient and personalized mobility. For civil engineers, the integration of connectivity and automation technologies into transportation infrastructure will be critical to facilitate the transition to smart and sustainable mobility.

Social Equity and Inclusive Access

As cities develop, social equity and inclusive access have become priorities within urban planning. In many cities, low-income areas lack adequate access to transportation services, limiting opportunities for employment, education, and health services for their residents (Pucher & Buehler, 2019). Urban planning should ensure that all communities, regardless of location and socioeconomic status, have access to affordable and accessible transportation options.

The concept of "spatial justice" in urban planning seeks to ensure that the benefits of infrastructure investments are distributed equitably and that the design of cities does not generate social exclusion (Gossling, 2021). For civil engineers, this implies designing infrastructure that responds to the needs of all inhabitants, including accessibility for people with disabilities and the creation of safe and accessible routes in all areas of the city.

The future of transportation and urban planning is marked by a series of challenges and opportunities that demand innovative, sustainable and inclusive solutions. Civil engineers have the responsibility to design infrastructures that not only respond to current mobility demands, but also adapt to a constantly changing context (Ruiz, 2024). From the electrification of transportation and autonomous mobility to the development of compact and accessible cities, success in urban and transportation planning will depend on the ability to anticipate future needs and implement resilient and sustainable solutions. Long-term vision and a commitment to equity, sustainability and efficiency will be essential to building cities that respond to the challenges of the 21st century and provide a better quality of life for all their inhabitants.

Chapter 2: Designing Sustainable Transportation Systems

Sustainability in transportation systems is not just an option, but a pressing need in the current context of climate change, population growth and rapid urbanization. Cities around the world face a reality in which mobility patterns must be radically transformed to reduce environmental impacts, improve air quality, and provide accessible and equitable transportation options for all citizens (UN-Habitat, 2023). This second chapter, entitled "Designing Sustainable Transportation Systems," invites readers to explore the principles, technologies, and strategies that enable the development of transportation systems that not only meet mobility needs, but also promote a balance between urban growth and environmental preservation. The design of a sustainable transportation system encompasses much more than the construction of infrastructure; it involves a comprehensive and multidimensional vision that considers emissions reduction, energy efficiency, accessibility, and resilience to future challenges. In this chapter, we will examine the innovations that are shaping the future of urban transportation, from electric transportation and micromobility to transit-oriented city planning. Each section highlights how civil engineers play a crucial role in creating transportation systems that minimize environmental impact and maximize social and economic benefits.

Why is sustainability in transportation systems essential? Transportation represents one of the main sources of greenhouse gas emissions in urban areas and, at the same time, is one of the most challenging sectors to decarbonize due to the complexity of the infrastructure and dependence on fossil fuels (International Energy Agency [IEA], 2022). In a context of growing concern about climate change, sustainability in transportation systems has become an essential component of urban and infrastructure policies, promoting solutions that reduce environmental impact and improve public health (Banister, 2018). Civil engineers are called to design systems that prioritize public transportation, clean energy, and infrastructure for active transportation modes, transforming mobility into a resource that promotes long-term sustainability.

Technologies and trends that are revolutionizing urban transportation The design of sustainable transportation systems is based on the adoption of technologies that enable greater efficiency and less dependence on fossil fuels. In this chapter, we explore emerging technologies such as electric public transport systems, which offer a clean and efficient alternative to traditional buses and trains. We also examine electric vehicle charging networks and their role in reducing emissions in private transportation. Micromobility, with options such as electric bicycles and scooters, is also presented as an innovative solution for short commutes, helping to reduce traffic and pollution in dense urban areas (Cohen & Shaheen, 2018).

People-centered design and accessibility for all

A truly sustainable transportation system must be inclusive and accessible to all inhabitants of a city, regardless of their location or socioeconomic status. This chapter addresses how transportation system design must consider equity and accessibility, ensuring that all communities have access to affordable and convenient mobility options. For civil engineers, this implies a people-centered approach, where transportation not only connects places, but also connects lives, opportunities, and essential services (Pucher & Buehler, 2019).

Resilience and preparedness for the future

Sustainability is not only a question of environmental impact, but also of resilience. As cities face extreme weather events, transportation systems must be designed to withstand and adapt to these changes, ensuring continuity of service and the safety of their users. In this chapter, we will explore how resilient infrastructure design, such as the integration of sustainable urban drainage and flood mitigation technologies, helps prepare cities for future climate challenges. This resilient approach enables civil engineers to build transportation systems that can withstand the test of time and the effects of climate change (United Nations, 2015).

A future of sustainable mobility

The transition to sustainable transportation is an opportunity to rethink our cities and build a future where mobility does not compromise the well-being of the planet or the quality of life of future generations. This chapter provides a practical and visionary guide for civil engineers and engineering students to understand the key elements for designing and planning transportation systems that meet the demands of the 21st century. From using clean technologies to creating more accessible and safer urban spaces, this chapter challenges readers to take a transformative approach to transportation engineering, leading the way to a future of sustainable mobility.

Dive into this chapter and discover how sustainable transportation systems can improve our cities, transform our relationship with the environment, and establish a legacy of resilience and equity. For civil engineers, designing sustainable transportation systems is not just a technical task; it is an ethical and professional commitment to the future of the planet and our communities.

2.1 Electric Public Transportation: Electric Trains, Trams and Buses

Electric public transport has become a key solution for cities seeking to reduce their carbon emissions and improve air quality. With increasing urbanization and growing mobility demands, electric transportation systems, such as electric trains, streetcars, and buses, offer a clean and efficient alternative to traditional transportation systems that rely on fossil fuels (International Energy Agency [IEA], 2022). This section examines the environmental, economic, and social benefits of electric public transport, as well as the challenges involved in its implementation and the technological innovations that are driving its development.

Emission Reduction and Air Quality Improvement

One of the greatest benefits of electric public transportation is its ability to significantly reduce greenhouse gas emissions and other pollutants that affect air quality in urban areas. According to Banister (2018), electric transportation systems, compared to fossil fuel-powered buses and trains, can reduce carbon emissions by up to 90%, which directly contributes to sustainability and public health goals in cities. The World Health Organization (2021) has warned that air pollution is one of the main causes of respiratory diseases in urban environments, and electric public transport is presented as an effective solution to mitigate these effects and improve the health of the population.

Electric trains and streetcars, in particular, are a high-capacity option that can reduce vehicular congestion by providing a fast and reliable means of transportation over long distances. In addition, by running on electricity, these systems generate less noise compared to combustion systems, which improves the quality of life in urban environments and creates quieter, more livable spaces (Gossling, 2021). Civil engineers play a crucial role in the design of these infrastructures, ensuring that electric transport networks are well integrated into the urban landscape and accessible to the population.

Operating Costs and Energy Efficiency

In the long term, electric public transport can also be more cost-effective due to its lower operating costs. Although the initial investment to establish electric grids and electric vehicles can be significant, the operating costs of electric trains, streetcars, and buses are much lower than those of their internal combustion counterparts, due to the reduced need for maintenance and the efficiency of electric motors (Rodrigue, 2020). According to a report by the International Energy Agency (IEA, 2022), electric motors convert approximately 85-90% of energy into motion, compared to 20-30% for internal combustion engines, which represents a considerable advantage in terms of energy efficiency.

These savings in operating costs can allow cities to redirect resources toward infrastructure improvements or even fare reductions, making public transportation more accessible to all. However, it is critical that civil engineers and urban planners collaborate to design electrical infrastructure to support this demand, including planning for charging stations and energy storage to avoid overloading the power grid.

Innovations in Battery and Charging Technology

One of the major challenges to the adoption of electric public transportation has been energy storage and charging technology. Innovations in lithium-ion batteries and other energy storage systems are enabling electric buses to operate for longer periods on a single charge, which improves the operational viability of these vehicles (Cohen & Shaheen, 2018). In addition, technologies such as fast charging and on-road charging systems allow vehicles to recharge their batteries at specific stations or even while in motion, increasing efficiency and reducing downtime.

Inductive charging systems, which enable wireless charging of electric vehicles using electromagnetic fields, are also beginning to be implemented in some cities, allowing electric buses and streetcars to be charged without the need for plugs or wires (Docherty et al., 2018). These technologies require advanced and well-planned infrastructure, underscoring the importance of civil engineering in the design and installation of energy charging and storage systems that are efficient, accessible, and safe.

Implementation Challenges and Urban Considerations

The implementation of electric public transport in urban areas poses certain logistical and financial challenges. Although the long-term benefits are significant, the initial investment to establish an electric transport network can be considerable and, in some cases, prohibitive for cities with limited budgets (Pucher & Buehler, 2019). In addition, retrofitting existing infrastructure and coordinating with the local power grid represent technical challenges that must be addressed with detailed planning and effective collaboration between civil engineers, governments, and energy providers.

Another important challenge is the integration of these systems into the urban space without negatively affecting the landscape or accessibility. Civil engineers must ensure that electric public transport infrastructure is well distributed and that charging systems are accessible and do not take up excessive public space, which could hinder other urban activities. When considering the design of charging stations and electric bus and streetcar routes, it is essential to plan in a way that minimizes disruption to traffic and maximizes accessibility for all users.

Electric public transport is a powerful solution for moving towards urban sustainability, offering significant benefits in terms of reduced emissions, operating costs and quality of life. However, its implementation requires meticulous planning, advanced infrastructure development and careful adaptation to the urban environment. For civil engineers, the design of electric public transport systems represents not only a technical challenge, but also an opportunity to contribute to building cleaner, more accessible and resilient cities. The transition to electric public transport is an investment in the future of our cities and in the health and well-being of their inhabitants.

2.2 Bicycle Paths and Non-Motorized Transportation Systems

In the quest for more sustainable, safe, and accessible cities, non-motorized transportation systems, such as bicycles and bikeway systems, have gained prominence in modern urban planning. Promoting active mobility not only reduces vehicular congestion and pollution, but also improves public health by encouraging a more active and healthy lifestyle (Pucher & Buehler, 2019). This section discusses the design, benefits, and challenges of implementing non-motorized transport systems in urban areas, highlighting their importance as part of a comprehensive approach towards sustainable urban mobility.

Benefits of Non-Motorized Transportation Systems

Non-motorized transport systems, such as bicycles and scooters, offer significant benefits for both users and the city as a whole. From an environmental perspective, the use of bicycles and other non-motorized modes reduces greenhouse gas emissions and decreases air pollution, which contributes to improving the environmental quality of urban areas (Banister, 2018). Active mobility also helps to decongest traffic, especially at peak hours, by providing an alternative to motorized vehicles over short distances.

At the individual level, non-motorized mobility improves physical and mental health. According to the World Health Organization (2021), activities such as cycling and walking can reduce the risk of cardiovascular disease and improve people's overall well-being. In addition, non-motorized transportation systems are affordable, making them a viable option for people of all socioeconomic levels. This makes non-motorized transport systems not only a sustainable option, but also an inclusive one, promoting social equity in cities (Litman, 2021).

Design and Infrastructure of Bicycle Paths

For non-motorized transport systems to function effectively, it is essential that cities have adequate and well-designed infrastructure. Bicycle paths, in particular, must be carefully planned to ensure the safety and comfort of users. Well-designed cycleways include protected lanes that separate cyclists from motorized traffic, adequate signage, and safe crossings that allow cyclists to ride safely (Gossling, 2021).

In addition, connectivity is a fundamental aspect of bikeway design. Bikeway networks must be well connected to each other and to other transport infrastructure, such as train or bus stations, to facilitate multimodal mobility. The creation of a network of interconnected bikeways allows users to move more efficiently and safely around the city, encouraging the use of bicycles as a means of daily transport. For civil engineers, this implies a design that not only responds to the needs of cyclists, but also integrates harmoniously into the urban environment, minimizing interference with vehicular and pedestrian traffic (Cohen & Shaheen, 2018).

Bicycle Sharing and Public Access Systems

Bike sharing systems have revolutionized urban mobility by making bicycles readily available to the general public. This type of system allows users to rent bicycles for a short period of time and return them at any station in the network, facilitating access to non-motorized mobility without the need to own a bicycle (Shaheen & Cohen, 2019). These systems are especially useful to complement public transport in the so-called "last mile", which connects users from transport stations to their final destination.

The success of bike sharing systems depends on well-planned infrastructure and efficient technology that allows users to locate and access bicycles quickly and conveniently. Civil engineers play a key role in the implementation of these systems, ensuring that bike sharing stations are located at strategic points and have sufficient capacity to meet demand. In addition, the infrastructure must be designed to withstand intensive use, minimizing vandalism and facilitating the maintenance of bikes and stations.

Implementation and Cultural Change Challenges

Although the benefits of non-motorized mobility are clear, its implementation in cities is not without challenges. One of the main obstacles is the car-based mobility culture that prevails in many cities and makes the use of bicycles and other non-motorized means less common. Changing this mentality requires awareness campaigns and infrastructure that ensures the safety of users, as many people are reluctant to use bicycles in cities with dense vehicular traffic (Pucher & Buehler, 2019).

In addition, the creation of infrastructure such as bicycle lanes and bike-sharing stations requires considerable investment, which can be a challenge in cities with limited resources. For civil engineers, this means designing non-motorized transport solutions that are economically viable and that maximize the use of urban space, integrating effectively into the environment and generating long-term benefits. At the regulatory level, it is also important that cities implement regulations and safety policies that protect the users of these systems and promote their responsible use.

Non-motorized transport systems and bikeways represent an effective strategy for moving towards more sustainable, equitable and healthy urban mobility . Through well-designed infrastructure and accessible bike-sharing systems, cities can reduce congestion and emissions, while encouraging more active and accessible living for all citizens. For civil engineers, designing these infrastructures involves not only a technical challenge, but also a commitment to community well-being and the creation of more inclusive and resilient cities. As urban mobility evolves, non-motorized transport systems are a key part of building a sustainable urban future.

2.3 Emission Reduction and Energy Efficiency in Transportation

In a world increasingly committed to sustainability and climate change mitigation, emissions reduction and energy efficiency in transportation have become key priorities. Since the transportation sector is a major source of greenhouse gas emissions, designing transportation systems that reduce dependence on fossil fuels and optimize energy use is essential to creating cleaner and healthier cities (International Energy Agency [IEA], 2022). This section explores strategies and technologies to reduce emissions and increase energy efficiency in transportation, highlighting the crucial role of civil engineers in implementing sustainable and resilient solutions.

Environmental Impact of Urban Transportation

Urban transportation has a significant impact on the environment. Emissions of carbon dioxide (CO2) and other pollutants, such as carbon monoxide and nitrogen oxide, are responsible for air pollution and contribute to global warming (Banister, 2018). In densely populated urban areas, the concentration of these pollutants directly affects public health, increasing the risk of respiratory and cardiovascular diseases among inhabitants (World Health Organization, 2021). Therefore, reducing emissions is fundamental not only for environmental sustainability, but also for the protection of public health.

For civil engineers, this challenge involves careful planning and the implementation of infrastructure that promotes the use of clean energy and reduces dependence on motorized transport. This includes the design of efficient public transport systems, the creation of low-emission zones, and the incorporation of supporting infrastructure for electric vehicles, such as charging stations (Gossling, 2021).

Electrification of Public and Private Transportation

Electrification is one of the most effective strategies to reduce emissions in transportation. Electric vehicles (EVs), both public and private, produce significantly fewer emissions than vehicles with internal combustion engines, especially if powered by electricity generated from renewable sources (Rodrigue, 2020). In public transport, the adoption of electric buses and electrified streetcars reduces the carbon footprint and improves energy efficiency, while private EVs represent a low-impact alternative for individual drivers (International Energy Agency [IEA], 2022).

The design and installation of charging stations is essential to support the transition to electrification. Fast-charging stations at strategic points in the city enable

electric vehicles to recharge their batteries in less time, facilitating their use both for long journeys and for daily urban transport. Civil engineers have the responsibility to plan and design this infrastructure, ensuring that it is well distributed and accessible in urban and suburban areas.

Promoting Active and Shared Mobility

In addition to electrification, promoting active mobility and ridesharing systems is another effective way to reduce emissions. Active mobility, such as cycling and walking, is not only zero-emission, but also improves public health and reduces traffic congestion (Pucher & Buehler, 2019). Cities that design adequate infrastructure for active mobility, such as bikeways and pedestrian areas, are helping to reduce the number of motorized vehicles on the streets, which lowers overall emissions and contributes to a healthier urban environment.

Ridesharing systems, such as car-sharing and ride-sharing, can also reduce emissions by reducing the number of private vehicles on the road. According to Cohen and Shaheen (2018), a single shared vehicle can replace 5 to 10 private cars, resulting in a significant decrease in emissions and energy use in urban transportation. For civil engineers, the challenge is to create an infrastructure that supports both active mobility and shared transportation, which implies the construction of bicycle stations, waiting areas for ride-sharing services, and parking spaces for shared vehicles.

Technological Innovation and Energy Efficiency

Technological innovation is key to improving energy efficiency in transportation. Advances in electric motors, long-life batteries and energy recovery systems are enabling vehicles to use less energy and generate fewer emissions. For example, regenerative braking systems in electric vehicles capture energy generated during braking and store it in batteries, increasing vehicle energy efficiency (Cohen & Shaheen, 2018).

Another significant innovation is the use of artificial intelligence (AI) and data analytics to optimize traffic flows and reduce fuel consumption. Real-time traffic management systems can adjust traffic lights and routes dynamically to reduce congestion and improve the efficiency of moving vehicles (Docherty et al., 2018). These technologies require advanced data and communication infrastructure, which represents a new field of opportunities for civil engineers in implementing intelligent and efficient transportation systems.

Reducing emissions and improving energy efficiency in transportation are essential components of urban sustainability. By electrifying transportation, promoting active and shared mobility, and adopting technological innovations, cities can reduce their carbon footprint and improve the quality of life of their inhabitants. For civil engineers, designing and implementing these solutions represents an opportunity to lead the way towards cleaner, more efficient and sustainable urban transportation. The transition to a low environmental impact transportation system is critical to the future of our cities and the planet.

2.4 Shared Mobility: Car-sharing, Ride-sharing and Micromobility

Shared mobility has emerged as one of the most innovative solutions to meet the challenges of congestion, pollution and efficiency in modern cities. This transportation model is based on the shared use of vehicles, from cars to bicycles and electric scooters, allowing citizens to access means of transportation without the need to own their own vehicle. This trend, driven by digital platforms and the demand for flexible mobility options, is transforming the way we get around, improving accessibility and reducing reliance on private vehicles (Cohen & Shaheen, 2018). This section explores the different types of shared mobility, their benefits and challenges, and the role of civil engineers in creating infrastructure that supports this model.

Car-sharing: An Alternative to the Private Car

Car-sharing allows users to rent vehicles by the hour or even by the minute, instead of having to purchase their own car. This model is ideal for people who need a vehicle only occasionally and are looking to avoid the costs and maintenance associated with car ownership. According to studies, each car-sharing vehicle can replace between 7 and 11 private cars on the road, significantly reducing the number of vehicles on the road and thus congestion and emissions (Shaheen & Cohen, 2019).

Civil engineers play a crucial role in the development of infrastructure that facilitates the use of car-sharing, such as the creation of dedicated parking lots and charging stations for shared electric vehicles. In addition, the integration of payment and digital management systems makes these services easy to use and accessible to the public. In the context of urban planning, car-sharing also helps to free up urban space that would otherwise be used for parking, allowing for a more efficient distribution of space in cities (Litman, 2021).

Ride-sharing: Efficiency and Emissions Reduction

Ride-sharing connects drivers who already plan to take a trip with other passengers who have the same or a similar destination. Platforms such as Uber and Lyft have popularized this model, allowing people to share rides and split costs. Ride-sharing not only offers a cost-effective and convenient alternative to the private car, but also reduces the overall number of vehicles on the road, which reduces congestion and emissions (Docherty et al., 2018).

In addition, ride-sharing improves energy efficiency in urban transport by maximizing the use of each vehicle and reducing empty kilometers traveled. However, for ride-sharing to have a positive impact on traffic and emissions, it is critical that it is well regulated and properly integrated into the urban transportation ecosystem. Civil engineers must work collaboratively with authorities to design infrastructure and regulations that support the use of ride-sharing in a safe and efficient manner, including waiting zones and designated pick-up and drop-off areas (Banister, 2018).

Micromobility: Shared Electric Bikes and Scooters

Micromobility, which includes the shared use of bicycles and electric scooters, has become a popular transportation option in modern cities. These lightweight, short-distance vehicles allow users to get around quickly and environmentally friendly, especially for last-mile or short-distance trips. Micromobility not only reduces congestion and emissions, but also promotes physical activity and improves accessibility in densely urbanized areas (Pucher & Buehler, 2019).

Civil engineers have the responsibility to design adequate infrastructure to support micromobility, including safe bicycle lanes, charging stations, and dedicated parking for bicycles and scooters. In addition, public space management must take into account the need to keep sidewalks and streets clear of obstructions, preventing micromobility vehicles from accumulating in areas of high pedestrian traffic. Careful planning and regulation are essential to integrate micromobility effectively and safely into the urban environment (Gossling, 2021).

Challenges and Opportunities in Shared Mobility

Although shared mobility offers multiple benefits, its implementation is not without challenges. One of the main obstacles is regulation, as the proliferation of shared vehicles, especially in the case of micromobility, can generate safety and congestion problems in pedestrian areas (Shaheen & Cohen, 2019). Cities must establish clear regulations that ensure the safe and orderly use of shared vehicles, and civil engineers must design infrastructure that minimizes negative impacts on public space.

Another challenge is equity of access. In many cities, shared mobility tends to be concentrated in central and high socioeconomic areas, limiting its accessibility for low-income or peripheral communities. To address this inequality, policymakers and urban planners should consider incentives and policies that promote the expansion of shared mobility in less accessible areas, ensuring that all citizens can benefit from these transportation alternatives (Cohen & Shaheen, 2018).

However, shared mobility also represents a great opportunity for sustainability and efficiency in urban transport. Supported by technological innovations and appropriate infrastructure, shared mobility can significantly reduce the number of vehicles on the road, improve energy efficiency and contribute to the creation of more livable and sustainable cities. For civil engineers, this involves not only the design of infrastructure, but also the implementation of integrated and sustainable solutions that optimize the use of urban space and reduce the environmental impact of transportation.

Shared mobility, in its various forms - car-sharing, ride-sharing and micro-mobility - is a key part of building a more efficient, accessible and sustainable urban transport system. Through well-planned infrastructures and appropriate regulations, civil engineers can support the adoption of these shared transport models, which not only optimize the use of resources, but also improve the quality of life in cities. Shared mobility represents a change in the way we understand transportation, challenging the dependence on the private vehicle and promoting a more flexible and environmentally friendly urban mobility.

2.5 Implementation of Intelligent and Accessible Payment Systems

In the context of modern urban mobility, the implementation of smart and accessible payment systems has become critical to improve the user experience and optimize the efficiency of transportation systems. These systems allow users to access multiple modes of transportation through integrated digital platforms, eliminating the need to carry cash and streamlining the payment process (Cohen & Shaheen, 2018). This section explores how smart payment systems contribute to the accessibility and efficiency of urban transportation, as well as the role of civil engineers in the planning and implementation of these technological infrastructures.

Facilitating Access to Multimodal Mobility

Smart payment systems enable seamless integration between different modes of transportation, facilitating the use of multimodal options such as rail, bus, bike-sharing, and ride-sharing in a single application (Litman, 2021). Mobility as a service (MaaS) allows users to plan, book, and pay for their trips through a single platform, making the transportation experience more efficient and accessible (Docherty et al., 2018). This integration lowers barriers to entry for users and encourages the use of more sustainable modes of transportation.

For civil engineers, the development of these platforms requires an adequate digital infrastructure that includes unified payment systems and robust data networks. This ensures that payment systems operate continuously and without interruption, even in densely populated areas or during peak hours. In addition, the implementation of multimodal platforms reduces the need for private cars, contributing to decongestion and reduced emissions in cities (Cohen & Shaheen, 2018).

Operational Efficiency and Cost Reduction

Smart payment systems also contribute to the operational efficiency of public transport systems. By digitizing the payment process, transportation systems reduce the amount of cash handled and minimize passenger waiting time. According to studies, the implementation of contactless payment systems, such as smart cards and mobile applications, has allowed waiting times in public transport to decrease by up to 30%, which optimizes the flow of users and reduces operating costs (Shaheen & Cohen, 2019).

In addition, these systems allow transportation agencies to collect real-time data on user behavior, which facilitates more accurate planning of routes, frequencies, and schedules (Banister, 2018). This analytical capability improves system efficiency and allows services to be adjusted according to demand, maximizing available resources. Civil engineers play an important role in the design and implementation of these systems, ensuring that payment infrastructures are safe, efficient and accessible to all users.

Accessibility and Social Inclusion

Accessibility is one of the major benefits of smart payment systems, as they allow all citizens, regardless of their geographic location or socioeconomic status, to have access to transportation services. Smart payment systems can be designed to offer reduced fares or discounts to specific groups, such as students, seniors, and low-income people, which promotes equity and social inclusion in urban transport (Gossling, 2021).

In addition, the digitization of payment reduces barriers for people who do not have cash or who prefer more secure and modern payment methods. However, the implementation of these systems must also take into account those who do not have access to smartphones or bank cards. For civil engineers , the challenge is to create inclusive payment solutions that enable the participation of the entire population, including alternative systems for those who do not have access to advanced technology or who prefer traditional payment methods.

Security and Privacy in Payment Systems

One of the key challenges in the implementation of smart payment systems is to ensure the security and privacy of user data. The digitization of payment involves the collection and management of large volumes of personal and financial data, which requires advanced security measures to prevent fraud and loss of information (Rodrigue, 2020). Civil engineers and urban planners must work together with cybersecurity specialists to develop payment infrastructures that protect user information and comply with privacy regulations.

In addition to security, privacy is also a growing concern among users of digital payment platforms. Collecting data on travel habits can be beneficial for urban planning, but it also poses risks in terms of individual privacy. Therefore, smart payment systems must be transparent and allow users to control what information they share and how it is used. Trust in the privacy and security of these systems is crucial to foster mass adoption and ensure that users are comfortable using these technologies (Cohen & Shaheen, 2018).

The implementation of smart and accessible payment systems is a powerful tool for transforming urban mobility, making it more efficient, accessible and adapted to users' needs. By enabling the integration of transport modes, reducing operating costs and improving accessibility, these systems represent a significant step towards more inclusive and sustainable urban transport. For civil engineers, the design of these technological infrastructures offers an opportunity to improve the mobility experience and contribute to the creation of more connected and safer cities. With a focus on safety, privacy and accessibility, smart payment systems are a key element in vision of an urban transport of the future, combining efficiency and equity.

Chapter 3: Urban and Land Use Planning

Urban planning is the backbone of any efficient, inclusive and sustainable city. Every space, every street and every building is part of an overall design that defines how people, services and the environment interact. In this context, planning and land use become fundamental tools for solving current urban problems and anticipating the future needs of cities. In a world that is rapidly moving towards urbanization, urban planning is no longer just about building physical infrastructure; it is a complex science that integrates social, economic, environmental and technological aspects, all aligned towards the goal of improving the quality of life of inhabitants. This chapter, entitled **"Urban Planning and Land Use,"** provides an immersion into the principles, strategies, and practices that civil engineers and urban planners can employ to create more livable and resilient cities.

The concept of urban planning ranges from spatial organization to resource allocation and infrastructure construction. At its core, urban planning seeks to optimize land use, balancing urban growth with the conservation of green spaces and the sustainable use of natural resources. However, modern planning also faces unprecedented challenges, such as the increasing density of urban populations, the need for sustainable and resilient cities in the face of climate change, and the inclusion of technology in urban spaces. For civil engineers and urban planners, urban planning is a constantly evolving field that demands continuous adaptation and innovative vision.

Comprehensive urban planning goes far beyond the construction of buildings and roads. Today, the discipline is deeply tied to sustainability, social justice and economic development. Well-planned cities provide accessible spaces for all, promote the development of connected communities, and create environments that are both livable and productive. Planning also has the potential to solve problems of social inequality by distributing resources and services equitably, providing access to transportation, education, health, and job opportunities to all inhabitants, regardless of their location within the city (Banister, 2018).

A well-planned city not only benefits its residents, but also boosts the economy by creating an attractive environment for business, tourism and investment. In addition, comprehensive planning ensures that resources are used efficiently, minimizing waste and environmental impact. From this perspective, civil engineers play a vital role in implementing solutions that maximize land use and improve connectivity and accessibility within cities.

Densification and the development of compact cities

One of the most relevant trends in urban planning today is densification and the development of compact cities. Instead of expanding to the peripheries, many cities are opting to grow "inward," increasing density in existing urban areas to maximize resources and reduce dependence on the automobile. This model, which includes concepts such as the "15-minute city" - where inhabitants have access to all essential services within a 15-minute radius on foot or by bicycle - allows for greater sustainability and land-use efficiency (Moreno et al., 2021).

The development of compact cities involves careful urban planning to avoid overloading infrastructure and services, and to ensure that densified areas have green and recreational spaces. The creation of compact cities also responds to a need for cultural change towards more sustainable lifestyles, encouraging active mobility and the use of shared spaces. For civil engineers, this means a focus on designing multifunctional infrastructures and maximizing resources to avoid uncontrolled sprawl and the problems associated with urban peripheries.

3.1 Principles of Sustainable Urban Planning

Sustainable urban planning is an approach that seeks to meet the needs of today's cities without compromising the ability of future generations to meet their own needs. In the context of rapid urban growth, climate change, and resource constraints, sustainability in urban planning has become a priority. This discipline integrates environmental, social, and economic factors into the design and management of urban spaces, creating a balanced environment that optimizes resources, reduces environmental impact, and improves the quality of life of inhabitants (Banister, 2018). This section explores the fundamental principles of sustainable urban planning and their application in designing cities that are resilient, inclusive, and efficient.

Efficient Use of Land and Control of Urban Growth

One of the central principles of sustainable urban planning is efficient land use. As cities grow, it is essential to avoid uncontrolled sprawl that consumes agricultural land, forests and natural areas. This phenomenon, known as "horizontal urbanization" or "urban sprawl," leads to a number of problems, such as increased transportation distances, fragmentation of natural habitat, and increased energy and water consumption (Litman, 2021). Instead of expanding to the peripheries, sustainable cities opt for controlled densification and "inward" development, promoting construction in existing urban areas to maximize the use of already established infrastructure and services.

The creation of compact cities, where residents can access basic services, transportation and employment without the need for long commutes, is a key strategy in urban sustainability. The idea of the "15-minute city" is an example of this approach, proposing that citizens can meet most of their daily needs within a short distance from home (Moreno et al., 2021). This model reduces car dependency and encourages active mobility, which, in turn, reduces emissions and traffic, improving the quality of life in the urban environment. For civil engineers, this implies designing multifunctional infrastructures that make the best use of urban space and promote safe and efficient coexistence between pedestrians, cyclists and vehicles.

Natural Resources and Green Areas Protection

The preservation of green areas and natural resources is another pillar of sustainable urban planning. Parks, gardens and nature reserves in cities not only provide spaces for recreation and leisure, but also contribute to improving air quality, regulating temperature and managing rainwater (Gossling, 2021). These green areas act as urban "lungs" that help mitigate the effects of climate change, reducing heat islands and absorbing polluting gases.

Sustainable planning also includes the protection of water bodies, wetlands and other natural resources critical to biodiversity and the environmental resilience of cities. The implementation of sustainable urban drainage systems (SUDS), such as rain gardens and retention ponds, allows for efficient water management, preventing flooding and improving water absorption in urban areas. For civil engineers, the challenge is to incorporate these green infrastructures into urban design, integrating them so that they not only fulfill an aesthetic function, but also provide long-term environmental and climate benefits (Rodrigue, 2020).

Promotion of Sustainable Mobility

Sustainable mobility is a key principle in urban planning, as transportation is one of the main sources of greenhouse gas emissions in cities. Fostering sustainable mobility involves creating infrastructures that promote the use of environmentally friendly modes of transport, such as public transport, cycling and pedestrian mobility (Cohen & Shaheen, 2018). Sustainable urban planning seeks to reduce dependence on the private automobile by promoting efficient and well-connected public transport networks that allow citizens to move around easily and without the need for personal vehicles.

Active mobility, such as cycling and walking, is also an essential component of sustainable mobility, as it not only reduces emissions, but also improves the health and well-being of the population. For civil engineers, designing safe bikeway networks, wide sidewalks and interconnected pedestrian areas is essential to encourage the use of active mobility. In addition, creating low-emission zones and implementing smart mobility technologies, such as digital payment systems and real-time traffic management, help reduce congestion and optimize the use of energy resources (Banister, 2018).

Inclusion and Equity in Access to Services and Resources

Social equity and inclusion are fundamental principles in sustainable urban planning. Cities must be accessible and offer equal opportunities to all their inhabitants, regardless of their geographic location, socioeconomic status or physical condition. This implies distributing services and resources equitably, ensuring that all areas of the city, including peripheries and low-income areas, have access to transportation, education, health, and recreational spaces (Pucher & Buehler, 2019).

For civil engineers, this means designing infrastructure that is accessible and safe for all, including ramps and access for people with disabilities, safe public spaces and inclusive recreation areas. Inclusive planning also means ensuring that urban development projects do not generate forced displacement or social exclusion, but rather contribute to community cohesion and the well-being of all citizens. Equity in urban planning is essential to create cities where all inhabitants can thrive and enjoy a good quality of life (Gossling, 2021).

The principles of sustainable urban planning are fundamental guidelines for meeting the challenges of urban growth in the 21st century. Through efficient land use, preservation of natural resources, promotion of sustainable mobility, and equity in access to services, urban planning can create urban environments that are livable, resilient, and environmentally responsible. For civil engineers, applying these principles means designing cities that not only meet technical requirements, but also contribute to the well-being and sustainability of communities. Sustainable urban planning is not just a strategy; it is a commitment to the future of our cities and to the quality of life for generations to come.

3.2 Urban, Suburban and Rural Areas: Characteristics and Connections

Urban planning focuses not only on the configuration of cities, but also on the harmonious integration between urban, suburban and rural areas. Each of these areas has unique characteristics and needs that must be considered in the context of city growth and development. Urban areas represent the centers of economic and social activity, while suburban and rural areas often play a complementary role, providing housing, resources and green space. However, the connections and planning of these areas are critical to creating a balanced and functional regional development system. This section explores the characteristics of each type of zone and discusses how integrated planning allows urban, suburban and rural areas to connect and benefit each other.

Characteristics and Functions of Urban Areas

Urban areas are the heart of economic, social and cultural activity. These areas, which are often highly dense, are home to most of the essential services and infrastructure, such as hospitals, schools, transportation networks and shopping centers. In urban areas, efficiency in the use of space is key, and design must maximize the functionality of each area to meet the demands of a large and diverse population (Banister, 2018). However, rapid urban growth presents significant challenges, such as congestion, pollution, and lack of green space.

Sustainable planning in urban areas focuses on optimizing land use, promoting densification and the development of compact cities. This includes the creation of multifunctional spaces and the integration of advanced technologies to improve resource management and urban mobility. For civil engineers, this involves designing efficient and sustainable infrastructures that enable a continuous flow of people and goods, improve connectivity and minimize the environmental impact of urban activities (Litman, 2021).

Suburban Areas: The Expansion of the City

Suburban areas, located on the fringes of urban areas, often combine urban and rural characteristics. Originally, suburban areas were developed as residential spaces for those seeking to escape the density and fast pace of cities, but at the same time wishing to be close to urban centers for access to employment and services. However, unplanned suburban growth has led to problems, such as uncontrolled urban sprawl, increased reliance on the automobile, and consumption of natural resources as cities expand horizontally (Gossling, 2021).

Planning for sustainable suburban areas requires an approach that balances the need for affordable housing and green space with proximity to essential services and the promotion of sustainable mobility. This includes the creation of public transport networks that effectively connect suburban areas to urban centers and reduce reliance on the automobile. In addition, the design of active mobility infrastructures, such as bicycle paths and pedestrian routes, helps to promote a healthy lifestyle and reduce emissions in these areas. Civil engineers play a key role in planning suburban areas that integrate sustainability and connectivity, ensuring development that does not compromise the natural environment or create an over-reliance on resources (Pucher & Buehler, 2019).

Rural Areas: Connection with the Natural Environment and Resources

Rural areas, although less dense, are essential to the well-being of urban and suburban areas, as they often provide food, water, energy and other natural resources. In addition, rural areas provide recreational and conservation spaces that are critical to ecological balance and tourism. However, rural areas often face problems such as lack of access to services, limited infrastructure, and depopulation due to migration to cities in search of employment and opportunities (Rodrigue, 2020).

To ensure an effective connection between rural and urban areas, planning should focus on improving transportation and infrastructure networks, facilitating access to basic services such as health, education and communication. The creation of rural-urban corridors allows products and services to flow efficiently between these areas, promoting balanced economic development and more equitable access to opportunities. Civil engineers have an essential role in designing sustainable roads and transportation networks that connect rural areas to urban centers while preserving the natural environment and minimizing environmental impact.

Connectivity and Integrated Planning between Zones

Connectivity between urban, suburban and rural areas is fundamental to sustainable regional development and social cohesion. Integrated planning allows each area to complement and support the others, creating an interdependent network that maximizes the benefits of each type of area. Urban areas rely on the resources of rural areas, while suburban areas act as bridges between these two spaces, offering a combination of urban infrastructure and natural environment (Cohen & Shaheen, 2018).

The development of transportation corridors and mobility networks connecting these areas is essential to reduce travel time, optimize the flow of goods and services, and reduce greenhouse gas emissions (Inter-American Development Bank [IDB], 2023). In addition, the implementation of buffer zones between urban and rural areas can help preserve biodiversity and limit uncontrolled urban sprawl. For civil engineers, integrated planning involves designing infrastructure that respects the specific characteristics and needs of each area, promoting balanced and sustainable development that benefits all inhabitants.

The planning of urban, suburban and rural areas should be understood as an interconnected system in which each area has a unique and complementary function. Creating a cohesive network that optimizes the characteristics and resources of each type of area is fundamental to the sustainable and balanced development of urban regions. For civil engineers, the challenge is to design infrastructure that facilitates connectivity and respects the particularities of each area, ensuring a balance between economic growth, environmental sustainability and social well-being. Integrated urban planning is a powerful tool for creating resilient and equitable communities, able to thrive in an increasingly interdependent environment .

3.3 Integration of Public Spaces and Green Areas

The integration of public spaces and green areas is a fundamental component of modern urban planning, as these spaces offer benefits that go beyond their aesthetic value. Green areas, such as parks, squares and gardens, are essential for the physical and mental well-being of inhabitants, and their presence improves the quality of life in cities. In addition, these public spaces contribute to sustainability by acting as "urban lungs," reducing heat islands, improving air quality, and serving as flood buffers. This section explores the multifaceted value of public spaces and green areas, and how their integration into cities fosters healthier, more resilient and inclusive communities.

Environmental Benefits of Green Areas

Green areas play a crucial role in mitigating the effects of climate change in cities. By providing shade and reducing surface temperatures, they help counteract the effect of urban heat islands, a phenomenon in which urbanized areas have significantly higher temperatures than their surroundings due to the concentration of asphalt and concrete (Gossling, 2021). In addition, plants and trees absorb carbon dioxide and release oxygen, improving air quality and helping to reduce greenhouse gas emissions (World Health Organization [WHO], 2021).

Natural drainage systems in green areas also help manage rainwater more efficiently, reducing the risk of flooding and promoting aquifer recharge. The implementation of rain gardens, green roofs and sustainable drainage systems allows cities to better manage rainfall and avoid overloading sewer systems (Rodrigue, 2020). For civil engineers, the design and implementation of these green infrastructures represents an opportunity to integrate sustainability into the urban environment, ensuring that public spaces not only beautify the city, but also contribute to its environmental resilience.

Public Spaces as Areas of Social Interaction and Well-being

Public spaces are meeting points that foster social interaction and create a sense of community. Squares, parks and recreational areas allow inhabitants to gather, participate in outdoor activities and develop social relationships, which is fundamental for social cohesion and mental wellbeing. According to WHO studies (2021), access to green areas and recreational spaces is related to a reduction in stress levels and an improvement in the mental health of urban dwellers.

The design of accessible and well-connected public spaces can also improve active mobility, encouraging people to walk or cycle instead of using motorized vehicles. This not only reduces congestion and emissions, but also promotes a healthier lifestyle. For civil engineers, creating these spaces means considering safety, accessibility, and functionality, ensuring that public spaces are inclusive and attractive to people of all ages and abilities (Pucher & Buehler, 2019).

3.3.3 Integration of Green Spaces in Urban Design

Modern planning seeks to integrate green spaces into the urban fabric so that they are accessible to all and form an integral part of people's daily lives. The idea of "green infrastructure" refers to a network of parks, gardens, green corridors and other natural spaces that connect different areas of the city, providing a continuous flow of nature in the urban environment (Banister, 2018). This integration allows residents to enjoy the benefits of green spaces without the need to travel long distances and improves connectivity between neighborhoods and communities.

A common practice in sustainable planning is the incorporation of green roofs and walls into buildings and other urban structures, which not only adds aesthetics, but also improves thermal insulation and air quality in densely built-up areas (Cohen & Shaheen, 2018). For civil engineers, this implies developing adapted infrastructures that can support these integrations and also maximize space utilization in environments where land is limited. The implementation of green infrastructure contributes to a more balanced urban design, where nature and architecture merge to create more livable and sustainable cities.

3.3.4 Contribution of Public Spaces to Social Equity

Public spaces and green areas play an important role in creating a more equitable city. Parks and plazas accessible to all allow people from different socioeconomic backgrounds to enjoy the same benefits, promoting inclusion and reducing inequalities. In many cities, however, the availability of green space is uneven, with a higher concentration in affluent areas and a lower presence in low-income communities (Litman, 2021). This disparity creates an urgent need to plan and distribute public spaces equitably.

Sustainable urban planning must ensure that all inhabitants have access to green areas and recreational spaces, regardless of their location. This not only improves the quality of life of citizens, but also helps to reduce social gaps. Civil engineers have a responsibility to develop projects that promote a fair distribution of these spaces, working in collaboration with local authorities to ensure that green areas and public spaces are inclusive and accessible to the entire community.

The integration of public spaces and green areas is essential for the development of sustainable, resilient and equitable cities. These spaces not only beautify the urban environment, but also offer environmental, social and health benefits that improve the quality of life of the inhabitants. For civil engineers, the design and implementation of public spaces and green areas represents an opportunity to contribute to a fairer city, where all inhabitants have access to the benefits of nature and can enjoy a healthy and vibrant living environment. The creation of an integrated network of green and public spaces is an investment in the well-being of present and future generations, and is a key component of the vision of a sustainable and livable city.

3.4 Compact City Models and Densification

The trend towards the development of compact cities and urban densification has gained importance in recent decades, driven by the desire to build sustainable, efficient and people-centered cities. Compact city models seek to optimize the use of land and resources, reducing dependence on the automobile and promoting the accessibility of services and public spaces within a well-defined area. In contrast to sprawl, which implies expansion towards the periphery, the compact city develops inward, promoting proximity, efficiency and equity in access to opportunities and services. This section explores the principles and benefits of the compact city and densification, as well as the challenges facing this strategy and the role of civil engineers in its implementation.

Principles of the Compact City Model

The compact city is based on a series of principles that seek to maximize land use and minimize environmental impact. One of the key concepts is the idea of the "15-minute city," which proposes that inhabitants can access most of their daily needs, such as stores, health centers, parks, and schools, within a 15-minute radius by walking or cycling (Moreno et al., 2021). This model encourages active mobility and reduces dependence on automobiles, which contributes to reducing carbon emissions and improving the quality of life of citizens.

Another fundamental principle is the mix of land uses, which consists of integrating housing, stores and services in the same area, creating neighborhoods that offer a complete living experience. This mix of uses not only promotes a more dynamic city, but also improves safety and social cohesion, as streets become more active and busy. For civil engineers, developing compact cities means designing multi-functional is and adaptable infrastructures that support the activity and constant flow of people, vehicles, and resources in a small space (Banister, 2018).

Benefits of Urban Densification

Urban densification offers multiple benefits for both inhabitants and the environment. In economic terms, it allows cities to optimize the use of infrastructure and services, reducing the costs of building and maintaining new infrastructure, such as transportation, water and electricity networks (Litman, 2021). The concentration of population in a limited area also facilitates the implementation of efficient and cost-effective public transport systems, as distances between destinations are shorter and demand is more constant.

From an environmental standpoint, urban densification contributes to reducing the expansion of the urban footprint and protecting rural and natural areas from uncontrolled urbanization. By encouraging the concentration of people and activities in specific areas, it limits the destruction of natural habitats and preserves natural resources. In addition, densification facilitates the creation of green areas and recreational spaces, which is essential for the mental and physical health of inhabitants (Gossling, 2021).

For civil engineers, this means designing infrastructure that supports higher user density without compromising functionality or the quality of the environment. This includes developing robust public transportation networks, efficient waste and water management systems, and creating mixed-use buildings that maximize the use of available space.

Challenges of Urban Densification

Despite its benefits, densification and urban compaction face several challenges. One of the most common problems is congestion, which can arise if infrastructure and services are not prepared to support the demand of a dense population. Compact areas require detailed planning to ensure that transportation, water, energy, and waste management systems are capable of meeting growing demand without becoming overburdened (Rodrigue, 2020).

Another challenge is the potential increase in the cost of housing, as higher demand in dense, well-connected areas can drive up prices, limiting access to housing for low-income people . This phenomenon, known as "gentrification," can lead to the displacement of local communities and create inequalities in access to the benefits of urban densification. To address this problem, it is important that densification policies include affordable housing strategies and mechanisms that protect vulnerable communities from real estate speculation (Cohen & Shaheen, 2018).

For civil engineers, the challenge is to design infrastructure and public spaces that mitigate these negative effects, ensuring that densification is carried out in a manner that is inclusive and respectful of the needs of all inhabitants. This may include creating accessible public spaces, implementing sustainable building standards, and integrating efficient transportation systems that avoid congestion.

The Role of Civil Engineers in the Compact City

Civil engineers play a key role in implementing the compact city model. From planning transportation systems to designing buildings and public spaces, their work is essential to creating cities that are both efficient, sustainable and livable. One of the biggest challenges for engineers is to maximize resource use and minimize environmental impact in dense environments, which involves innovating in the design and construction of multifunctional and sustainable infrastructure (Pucher & Buehler, 2019).

The creation of mixed-use buildings, which combine residential, office, retail and recreational areas in one location, is a common strategy in the compact city and requires careful planning to ensure that these structures are safe, accessible and efficient in their use of energy and resources. In addition, civil engineers are responsible for developing sustainable mobility systems, such as interconnected public transportation networks, bicycle paths and pedestrian zones, that facilitate movement within these compact areas without relying on the automobile.

The implementation of technologies such as renewable energy, water recycling and smart waste management systems also plays a crucial role in the compact city. These innovations enable cities to operate more efficiently and reduce their ecological footprint. The ability of civil engineers to integrate these technologies and adapt to the demands of dense environments is critical to the success of compact cities in the future.

The compact city model and urban densification offer a vision of more sustainable, efficient and accessible cities, where services and opportunities are available to all inhabitants. However, their implementation requires meticulous planning and a commitment to address challenges that may arise, such as congestion, housing affordability, and gentrification. For civil engineers, building compact cities is an opportunity to lead in designing urban environments that respond to the needs of the 21st century, providing infrastructure that improves quality of life and reduces environmental impact.

The development of compact cities is not only an urban planning strategy, but also a commitment to a future in which cities can thrive in a sustainable and equitable manner. With a combination of innovation, technology and social engagement, civil engineers can help shape cities that are not only efficient and functional, but also inclusive and resilient.

3.5 Transit-Oriented Development (TOD): Linking Transportation and Land Use

Transit-Oriented Development (TOD) is an urban planning strategy that seeks to integrate transportation and land use to create efficient, accessible and sustainable urban environments. This model focuses on creating high-density areas around public transport stations, facilitating access to services and reducing the need to travel by car. TOD not only improves mobility and reduces congestion, but also promotes compact urban development that optimizes land use and supports the creation of more connected communities. This section explores the principles and benefits of TOD, the challenges of its implementation, and the role of civil engineers in building infrastructure that enables its success.

Fundamental Principles of Transportation-Oriented Development

TOD is based on several key principles that seek to encourage the use of public transportation and active mobility. One of the main pillars is proximity, i.e., the design of urban spaces in which essential services and destinations, such as stores, schools and recreational areas, are within easy reach of residents within a radius close to transit stations (Cervero & Murakami, 2009). This reduces car dependency and facilitates access to public transportation, making it more convenient and attractive to residents.

Another fundamental principle of TOD is the mix of land uses. By combining residential, office, retail, and recreational space in areas near transit stations, TOD creates dynamic, vibrant neighborhoods that promote social interaction and economic activity. For civil engineers, this approach means designing multifunctional and accessible infrastructure that allows residents to take advantage of public transportation and pedestrian connections in a seamless and unobstructed manner (Litman, 2021).

Benefits of TOD for Mobility and the Environment

Transit-Oriented Development offers multiple benefits in terms of mobility and sustainability. By reducing the need to travel by automobile, TOD helps reduce greenhouse gas emissions and air pollution, contributing to climate change mitigation (Banister, 2018). In addition, TOD improves the efficiency of public transportation by increasing demand and justifying investments in high-capacity systems, such as rail and bus rapid transit, which benefits residents by providing them with a reliable and accessible mobility alternative.

From a social standpoint, TOD fosters the creation of more connected and cohesive communities. By clustering housing, stores and services around transportation hubs, residents can enjoy greater proximity to their destinations, which promotes social interaction and strengthens the sense of community. For civil engineers, this involves designing infrastructure that promotes active mobility, such as wide sidewalks, safe bicycle paths and accessible crosswalks, creating urban environments that encourage the use of public transportation and physical activity (Cervero & Murakami, 2009).

Challenges of Transportation-Oriented Development

Despite its many benefits, TOD faces several challenges that must be addressed for successful implementation. One of the most common problems is resistance to change, as in some cities residents are accustomed to relying on cars and may be reluctant to adopt a public transport-oriented lifestyle. This paradigm shift requires effective public policies and awareness campaigns that inform citizens about the benefits of TOD for mobility, the environment and quality of life (Rodrigue, 2020).

Another challenge is the initial cost of public transport infrastructure, especially in areas where the existing system is limited. Investment in high-capacity transportation networks and construction of accessible stations can be high, which can limit TOD implementation in low-income areas or in cities with limited resources. To overcome this obstacle, civil engineers and planners must work together with the public and private sector to develop financing models that make the development of these essential infrastructures possible (Cohen & Shaheen, 2018).

In addition, gentrification is a potential problem in areas developed under the TOD model. As transportation access and quality of life improve in certain areas, housing prices may increase, displacing low-income residents and limiting access to TOD benefits for all. This phenomenon can be addressed through affordable housing policies and subsidies that ensure that local communities are not displaced by transit-oriented development (Pucher & Buehler, 2019).

Role of Civil Engineers in Transportation-Oriented Development

Civil engineers play a crucial role in the implementation of TOD, as their work is fundamental to building the transportation infrastructure and services that enable the success of this model. This includes the planning and design of accessible and safe transportation stations , as well as the creation of active mobility networks, such as bikeways, pedestrian routes, and interconnection areas that facilitate access to public transportation (Inter-American Development Bank [IDB], 2023). In addition, civil engineers must ensure that infrastructure is designed to support the constant flow of users and that it meets the needs of all inhabitants, including people with disabilities.

Another important aspect is the incorporation of smart mobility technologies, such as traffic management systems and digital payment platforms, which facilitate the use of public transport and optimize the flow of people and vehicles in high-density areas (Litman, 2021). These technologies allow cities to manage mobility more efficiently and enable inhabitants to access transportation conveniently and without interruption.

Transit-Oriented Development is a transformative strategy for creating more sustainable, accessible and people-centered cities. Through proximity, land use mix, and the integration of transportation and active mobility infrastructure, TOD fosters a way of life in which public transportation and active mobility are viable and attractive options for citizens. However, for TOD to succeed, it is critical to address associated challenges such as cultural resistance, infrastructure cost, and risk of gentrification.

For civil engineers, transportation-oriented development represents an opportunity to lead the way in creating infrastructure that meets the demands of the future and improves the quality of life in cities. Through careful planning and the implementation of innovative technologies, engineers can contribute to creating an urban environment in which transportation is efficient, equitable, and sustainable. TOD is a step towards more connected and resilient cities, where all inhabitants can enjoy an accessible and quality living environment.

Chapter 4: Intelligent Technologies and Sustainability in Urban Transportation

Urban transportation is at a crossroads: on the one hand, it faces the challenges of a growing urban population, traffic congestion and pollution; on the other, it has the opportunity to transform itself thanks to the technological revolution and the demand for sustainability. In a world that is increasingly connected and conscious of its environmental footprint, smart technologies offer innovative solutions that can profoundly change the way people move within cities. This fourth chapter, entitled **"Smart Technologies and Sustainability in Urban Transportation,"** explores how technology is revolutionizing urban transportation and enabling a more sustainable and efficient approach, as well as the role that civil engineers play in this transformation.

Smart technologies not only optimize existing transportation systems, but also create new possibilities for the development of infrastructure and services that meet the needs of the 21st century. From autonomous transportation systems to mobility as a service (MaaS) and intelligent traffic management, this chapter explores how these innovations are redefining urban transportation, promoting smoother, cleaner, and more user-centered mobility. For civil engineers and urban planners, the integration of smart technologies is an opportunity to lead the way in creating more livable and connected cities, where transportation is accessible, equitable, and environmentally responsible.

Smart technologies at the service of urban mobility

Technology has changed the way we interact with transportation, from planning our trips to paying for services. Today, citizens can plan their trips and access multiple transportation options through a single app, facilitating multimodal mobility and reducing reliance on the private automobile (Cohen & Shaheen, 2018). Mobility as a service (MaaS) is one of the most promising emerging concepts, which integrates diverse transportation options-such as buses, trains, bikes, and car sharing-into a digital platform that allows users to plan, book, and pay for their trips from a single location. This technology-enabled approach facilitates access to sustainable transportation options and makes the use of public transportation and active mobility more convenient and attractive.In addition, artificial intelligence (AI) and real-time data analytics are revolutionizing traffic management and energy efficiency in transportation. Intelligent traffic management systems use real-time data to optimize traffic light timing, adjust bus and train routes, and reduce waiting times, which improves user experience and reduces emissions generated by vehicular congestion (Docherty et al., 2018). For civil engineers, the implementation of these technologies requires advanced and adaptive infrastructure that allows cities to manage traffic dynamically and efficiently, improving the flow of people and goods and minimizing environmental impact.

Sustainability in transportation: an urgent need

Sustainability is an essential objective in the development of modern urban transport. Cities are responsible for a large share of global emissions of CO2 and other pollutants, and transportation accounts for a significant portion of these emissions. The transition to more sustainable transportation is critical to reduce the environmental footprint of cities and to protect the health and well-being of citizens (Banister, 2018). In this chapter, we will discuss how transportation electrification, active mobility, and renewable energy sources are enabling cities to move towards cleaner and healthier mobility.

Electrification of public transportation, for example, has proven to be one of the most effective strategies for reducing emissions in densely populated urban areas. Electric buses, electrified commuter trains and streetcar systems not only reduce greenhouse gas emissions, but also reduce noise pollution, creating a more pleasant urban environment In addition, integrating renewable energy sources into urban transport systems allows cities to reduce their dependence on fossil fuels and become more resilient to energy price fluctuations (International Energy Agency [IEA], 2022).

The role of civil engineers in smart and sustainable mobility.

The transformation of urban transport towards smarter and more sustainable models places civil engineers at the center of this evolution. Their role goes far beyond the construction of physical infrastructure; it includes the integration of advanced technologies and the planning of complex systems that facilitate smooth and green mobility (World Economic Forum, 2022). For engineers, this transition requires constant adaptation and learning of new tools and methodologies, as well as close collaboration with technology experts, local authorities and the private sector to ensure that urban development meets the needs of all citizens.

The planning and implementation of these innovations are not without challenges. The adoption of smart and sustainable technologies requires significant investments and, in some cases, a change in the mindset of both professionals and users. In addition, it is critical to ensure that the use of smart technologies in transportation does not increase social gaps, but rather promotes accessible and equitable mobility for all citizens. Civil engineers, in collaboration with other key stakeholders, must work to develop transportation infrastructures that not only respond to current needs, but also adapt to future changes and promote equity in access to mobility.

Smart and sustainable urban transportation is not just a trend; it is an urgent need and a unique opportunity to transform cities into healthier, more connected and resilient environments. Smart technologies offer powerful tools to optimize urban transportation, while sustainable solutions enable cities to reduce their environmental impact and improve the quality of life of their inhabitants. This chapter invites civil engineers, urban planners, and policymakers to see technology and sustainability as allies in creating urban transportation that meets the challenges of the present and future.

As you explore this chapter, you will discover how the convergence of technology and sustainability is shaping a new era of urban mobility. From the electrification of transportation to mobility-as-a-service and autonomous vehicles , technology offers endless possibilities for creating cities where transportation is more efficient, accessible, and environmentally friendly. For civil engineers, this is an opportunity to lead the shift to a future where cities are not only functional, but also sustainable and just.

4.1 Electrification of Urban Transportation: Buses, Trains and Freight Networks

Electrification of urban transport is one of the most promising strategies to reduce the carbon footprint of cities and improve air quality in densely populated environments. Transitioning to electric transportation systems, such as buses, trains, and streetcars, allows for lower emissions of greenhouse gases and other pollutants, which contributes significantly to sustainability and environmental protection goals (International Energy Agency [IEA], 2022). This section explores the benefits and challenges of urban transport electrification, as well as the crucial role of civil engineers in the implementation of charging infrastructure and in the design of efficient and accessible electric transport systems.

Environmental and Social Advantages of Electrification

Electric transportation reduces greenhouse gas emissions and air pollution in cities. Compared to transportation systems powered by combustion engines, electric buses and trains emit considerably less carbon dioxide and do not produce particulate pollutants, thus improving air quality in urban environments and reducing the risk of respiratory diseases among inhabitants (Gossling, 2021). The World Health Organization (2021) has highlighted the importance of reducing air pollution in cities, noting that exposure to pollutants is associated with an increase in diseases such as asthma and other respiratory conditions.

In addition to their environmental benefits, electric vehicles contribute to reducing noise pollution, as their engines are significantly quieter than combustion engines. This improves the quality of life in urban areas, especially in densely populated residential and commercial areas. For civil engineers, the electrification of transportation represents an opportunity to develop urban infrastructures that not only respond to mobility needs, but also contribute to a healthier and more environmentally friendly living environment (Banister, 2018).

Load Network Infrastructure and its Distribution in the Urban Space

One of the biggest challenges in the electrification of urban transportation is the creation of an adequate charging infrastructure to enable the efficient operation of electric vehicles. Strategic planning and distribution of charging stations are essential to ensure that buses, trains and other electric vehicles can operate continuously without interruption. Fast charging stations, which allow batteries to be recharged in minutes rather than hours, are especially important for public transport, as they reduce downtime and increase operational efficiency (Rodrigue, 2020).

In dense urban areas, space for charging stations is limited, which presents a challenge for civil engineers. It is essential to design charging networks that integrate harmoniously into the urban environment, minimizing visual impact and maintaining accessibility for pedestrians and other users of public space. The implementation of charging stations in strategic locations, such as bus terminals and train stations, optimizes the use of resources and facilitates access to charging infrastructure for public transport operators. In addition, wireless charging technology and on-road charging systems, such as inductive charging lanes, offer innovative solutions that can improve the efficiency of charging infrastructure and reduce the need for physical space (Cohen & Shaheen, 2018).

Advances in Battery Technology and Renewable Energies

Battery technology has evolved considerably in recent years, making it possible for electric vehicles to have an ever-increasing range and a longer service life. Lithium-ion batteries are the most common today, although new technologies are being developed, such as solid-state batteries, which promise to offer greater energy efficiency and shorter charging times. These advances are crucial to the success of transport electrification, as allow electric vehicles to travel longer distances without recharging, which is particularly beneficial for long-distance public transport routes (International Energy Agency [IEA], 2022).

In addition, integrating renewable energy into the charging network is critical to maximizing the environmental benefits of electrification. By using energy sources such as solar and wind, cities can further reduce carbon emissions and ensure that the electricity used to charge vehicles is clean and sustainable. For civil engineers, this involves designing charging infrastructure that can accommodate diverse energy sources and maximize the use of renewable energy, which requires careful planning and close collaboration with the energy sector (Litman, 2021).

Challenges in the Implementation and Maintenance of Electric Transportation Systems

The electrification of urban transport presents several logistical and technical challenges. One of the main ones is the initial cost of charging infrastructure and electric vehicles, which can be significantly higher than that of conventional vehicles. Although the operating costs of electric vehicles tend to be lower due to their lower energy consumption and maintenance, the initial investment remains an obstacle for many cities, especially those with limited budgets (Cervero & Murakami, 2009).

Another challenge is the maintenance and management of the charging infrastructure and power grids. Electricity demand in a densely populated city can be considerably high, especially at peak times, requiring an efficient energy management system that avoids overloading the grid. For civil engineers, this means developing solutions that optimize energy use and minimize wear and tear on the charging infrastructure, ensuring that electric transportation systems are sustainable and functional in the long term.

In addition, training and technical skills development for transportation and maintenance personnel are crucial for the adoption of electric transportation systems. The technology of electric vehicles is different from that of combustion vehicles, and maintenance teams need to have the right skills to operate and repair these vehicles safely and efficiently. For civil engineers, this also means working collaboratively with the education and professional sector to ensure that workers have the necessary skills to support the transition to electric transportation (Gossling, 2021).

The electrification of urban transport represents a crucial step towards sustainability and improved quality of life in cities. By reducing emissions and pollution, electric transport systems contribute to a cleaner and healthier urban environment, promoting mobility that is both efficient and environmentally friendly. For civil engineers, the electrification of transportation involves designing

of advanced and adaptive infrastructures that respond to the demands of a more sustainable and accessible mobility.

This shift towards electrification is not only an opportunity to improve urban transportation, but also a responsibility to ensure that cities adapt to the challenges of the future. Through the implementation of innovative charging stations, the development of efficient battery technologies and the use of renewable energy, civil engineers are at the center of this transformation, leading the way towards urban transportation that responds to the sustainability and resilience needs of our era.

4.2 Mobility as a Service (MaaS): Integration and Flexibility in Urban Transportation

Mobility as a Service (MaaS) is an innovative urban transportation strategy that integrates multiple mobility options into a single digital platform, offering users the ability to plan, book, and pay for their trips from an app or website. This model transforms the way citizens interact with transportation, giving them access to options such as public transit, bike-sharing, ride-sharing vehicles, and electric scooters in one place (Cohen & Shaheen, 2018). MaaS promotes more flexible, accessible, and sustainable mobility by facilitating the adoption of environmentally friendly means of transportation and reducing reliance on the private automobile. This section explores the benefits, challenges, and principles of MaaS, and the role of civil engineers in the infrastructure and technology that make this model possible.

MaaS Principles and Advantages

The MaaS model is based on simplicity and accessibility. By combining various transportation options in a single application, MaaS allows users to organize their journeys in a more efficient and personalized way. One of the key principles of MaaS is the integration of transportation modes into a single digital platform, which not only simplifies the trip planning process, but also allows users to compare options and choose the one that best suits their needs in terms of time, cost, and sustainability (Docherty et al., 2018).

MaaS offers multiple advantages for cities and their inhabitants. For citizens, this model facilitates access to sustainable transportation options, promoting the use of public transport, bicycles and other means of shared transport, which helps reduce congestion and greenhouse gas emissions. For civil engineers and urban planners, MaaS represents an opportunity to design more livable and connected cities, where mobility services are available to all inhabitants and are efficiently integrated into the urban environment (Gossling, 2021).

MaaS Infrastructure and the Role of Civil Engineers

The implementation of MaaS requires advanced technological infrastructure and robust connectivity to enable citizens to access transportation services in real time. Civil engineers play an essential role in designing physical and digital infrastructure to support this model. This includes creating interconnection points where users can switch from one mode of transport to another seamlessly, such as train stations integrated with bike-sharing terminals or bus stops interconnected with electric scooter stations.

In addition, data infrastructure is critical to the success of MaaS. Engineers must collaborate with software developers and transportation operators to create systems that collect, manage, and share real-time information on the availability and wait times of each transportation option. This data enables users to make informed decisions and improves the efficiency of the transportation system as a whole. Also, digital payment infrastructure allows users to pay for multiple transportation services in a single transaction, which facilitates adoption and use (Litman, 2021).

Sustainability and Reduction of Automobile Dependence

One of the greatest benefits of MaaS is its potential to reduce dependence on the private automobile, promoting more sustainable and efficient modes of transportation. By offering a variety of transportation options on a single platform, MaaS facilitates multimodal mobility, allowing users to combine different modes of transportation according to their specific needs. For example, a user can take a train for most of their journey and, upon arrival at the final station, use a shared bicycle or scooter to travel the last mile. This mobility model reduces the need for private car use, which in turn reduces congestion and emissions (Banister, 2018).

In addition, MaaS encourages active mobility and the use of low-emission modes of transportation. By integrating options such as bike sharing and electric scooters, cities can reduce the carbon footprint of urban commuting and improve public health by promoting a more active lifestyle. For civil engineers, this approach requires the planning of active mobility infrastructures, such as bikeways and pedestrian zones, that allow a smooth transition between different modes of transport and ensure the safety of all users (Pucher & Buehler, 2019).

MaaS Implementation and Adoption Challenges

Despite its many benefits, MaaS implementation faces several challenges. One of the main obstacles is fragmentation among transport operators. In many cities, public transport services, bike sharing and other mobility modes are managed by different entities, making it difficult to integrate all these services into a single platform (Rodrigue, 2020). Collaboration between different actors in the transport sector is essential to overcome this fragmentation and enable a seamless and seamless user experience.

Another important challenge is accessibility and equity in access to MaaS. For this model to be truly inclusive, it is essential that all citizens, regardless of their socioeconomic status, can access integrated mobility services. This involves considering affordable payment options, such as reduced fares for students and low-income individuals, and ensuring that mobility services are available in all areas of the city, including peripheral and low-income areas. Civil engineers and urban planners have a responsibility to design infrastructure that ensures that mobility services are equitably distributed and accessible to all (Cohen & Shaheen, 2018).

Mobility as a Service (MaaS) is a transformative strategy that can revolutionize the way citizens interact with urban transportation, promoting a multimodal, flexible and sustainable approach. By integrating various modes of transportation into a single platform, MaaS facilitates access to efficient and environmentally friendly mobility options, reducing dependence on the private automobile and improving the quality of life in cities. For civil engineers, the implementation of MaaS represents an opportunity to innovate in the design of physical and digital infrastructures that support this model and promote inclusive and environmentally friendly mobility.

However, MaaS is not without its challenges. Its success depends on close collaboration between transport operators, local authorities and technology experts, as well as adequate infrastructure that allows users to easily access integrated, real-time transport services. As cities evolve towards a more sustainable future, MaaS presents itself as a powerful tool for transforming urban transport into a system that is both efficient and accessible, offering citizens the freedom to choose the mobility option that best suits their needs while reducing their impact on the planet.

4.3 Autonomous and Connected Vehicles: Innovation and Challenges in Urban Transportation.

The advent of autonomous and connected vehicles represents one of the most revolutionary developments in the history of urban transportation. These vehicles, equipped with advanced sensor technology, artificial intelligence (AI) and real-time connectivity, have the potential to profoundly transform the way people move around the city, as well as change the structure and design of urban infrastructure. Autonomous vehicles promise to improve road safety, reduce congestion, and optimize the use of energy resources, while connected vehicles enable constant interaction with other mobility systems, providing valuable data for efficient traffic management (Docherty et al., 2018). This section explores the benefits, challenges, and principles of autonomous and connected vehicles in the context of urban transportation and discusses the role of civil engineers in adapting infrastructures to facilitate their integration.

Advantages of Autonomous Vehicles in Urban Mobility

Autonomous vehicles offer multiple benefits in terms of safety and efficiency. By eliminating human intervention in driving, these vehicles have the potential to drastically reduce traffic accidents, many of which are caused by human error (Banister, 2018). Autonomous vehicles are designed to react in fractions of a second to unforeseen situations, which improves the safety of all road users. In addition, the technology in these vehicles allows them to maintain a safe distance from other cars and to adapt their speed and routes according to traffic conditions, which optimizes vehicular flow and reduces traffic jams in urban areas.

Autonomous vehicles also offer the opportunity to improve the accessibility of urban transportation for people with reduced mobility, the elderly, and those without a driver's license. By providing efficient and accessible transportation, these vehicles allow more people to move around the city without relying on traditional private or public transportation (Cohen & Shaheen, 2018). For civil engineers, implementing autonomous vehicles means designing adapted infrastructure, such as dedicated lanes and charging stations for electric vehicles, to facilitate their safe and efficient operation in the urban environment.

Connected Vehicles and Intelligent Traffic Management

Connected vehicles are an extension of autonomous vehicles, with the ability to communicate in real time with other vehicles, infrastructure and traffic control systems. This connectivity enables more efficient traffic management, as vehicles can receive real-time information on road conditions, weather conditions, and potential incidents, allowing them to make informed decisions that optimize vehicle flow and reduce congestion (Litman, 2021). V2V (vehicle-to-vehicle) and V2I (vehicle-to-infrastructure) communication systems are key technologies in the operation of connected vehicles, enabling constant synchronization between all elements of the transportation system.

The ability of connected vehicles to exchange information with intelligent traffic lights, traffic sensors and route management systems helps minimize waiting time and improves energy efficiency by reducing fuel consumption and emissions (Rodrigue, 2020). For civil engineers, this advancement involves designing and adapting infrastructure with sensors and communication devices that facilitate connectivity and enable real-time data collection and analysis. This smart infrastructure is essential to maximize the benefits of connected vehicles and ensure smoother and more sustainable urban mobility.

Autonomous and Connected Vehicle Deployment Challenges

Although the benefits of autonomous and connected vehicles are obvious, their implementation presents multiple technical, ethical and regulatory challenges. One of the main issues is the need for adapted infrastructures, as current roads and urban streets are not always equipped to support the operation of autonomous vehicles. The lack of digital traffic signals, intelligent traffic lights and V2I communication systems can limit the effectiveness of these vehicles and their ability to operate safely (Gossling, 2021).

Another challenge is data security and privacy. Autonomous and connected vehicles collect a large amount of data about users and the environment, raising concerns about the protection of personal information and the risk of cyberattacks. The possibility that vehicle control systems could be breached is a threat that needs to be addressed through strict regulations and advanced security systems. For civil engineers, this aspect represents an additional challenge, as transportation infrastructure must be designed with security measures that prevent unauthorized access to vehicle control systems and data (Docherty et al., 2018).

In addition, public acceptance of autonomous vehicles is also a challenge, as the transition to automated driving can generate uncertainty and resistance among users. It is critical that engineers and planners work together with the public sector and communities to raise awareness of the benefits and safety aspects of autonomous and connected vehicles, encouraging gradual and reliable adoption.

The Role of Civil Engineers in the Integration of Autonomous and Connected Vehicles.

Civil engineers are key players in the integration of autonomous and connected vehicles into the urban transportation system. Their responsibility includes the design and construction of infrastructures that enable a safe transition to autonomous mobility, such as the creation of dedicated lanes for autonomous vehicles, the installation of intelligent traffic lights, and the implementation of charging stations for electric vehicles. The planning of these infrastructures must not only consider technical aspects, but also safety and accessibility for all users.

Real-time data collection and analysis is another crucial aspect of the role of civil engineers in the implementation of autonomous and connected vehicles. The data generated by these vehicles can provide valuable information for urban planning and infrastructure improvement. Civil engineers must work in collaboration with data analytics specialists and transportation authorities to develop systems that use this information effectively, optimizing vehicle flow and improving the quality of life in cities.

Sustainability also plays an important role in the design of infrastructure for autonomous vehicles. Civil engineers can contribute to reducing the carbon footprint of urban transportation by integrating charging stations for electric vehicles and designing traffic management systems that reduce emissions. These efforts contribute to a greener and more efficient transportation model that meets the sustainability demands of modern cities.

The advent of autonomous and connected vehicles marks the beginning of a new era in urban transportation, with technology and connectivity redefining the way people get around and experience the city. These vehicles offer the opportunity to improve safety, reduce congestion, and make transportation more accessible to all. However, their implementation requires advanced infrastructure, data security systems, and gradual acceptance by citizens.

For civil engineers, the integration of autonomous and connected vehicles is an exciting challenge and an opportunity to innovate in the design of urban infrastructure that meets the needs of the 21st century. Through careful planning, the use of advanced technologies and collaboration with other sectors, engineers can help build safer, more connected and sustainable cities, where autonomous and connected mobility is part of the urban transportation of the future.

4.4 Intelligent Traffic Management: Optimization and Efficiency in Urban Mobility

Intelligent traffic management is a fundamental tool for optimizing the flow of vehicles in cities and reducing congestion, emissions and travel time. Thanks to technological advances, it is now possible to use artificial intelligence (AI) systems, sensors, cameras and real-time data analysis to manage traffic efficiently and in response to changing road conditions. These intelligent systems enable not only dynamic traffic light synchronization, but also optimal route planning, real-time incident identification, and public transport prioritization (Docherty et al., 2018). This section explores how intelligent traffic management is transforming urban mobility, its benefits, challenges, and the role of civil engineers in its implementation.

Key Technologies in Intelligent Traffic Management

Intelligent traffic management is based on the implementation of advanced technologies that enable real-time data collection, processing and analysis. These technologies include traffic sensors, surveillance cameras, and V2I (vehicle-to-infrastructure) and V2V (vehicle-to-vehicle) communication systems, which allow vehicles and traffic infrastructures to communicate with each other to improve efficiency and safety on the roads (Rodrigue, 2020). These technologies collect data on traffic volume, vehicle speed, and road conditions, which are used to adjust traffic lights and optimize vehicle routes.

Artificial intelligence (AI) also plays a crucial role in intelligent traffic management by analyzing large volumes of data in real time and making instant decisions to improve vehicle flow. AI systems can identify patterns in traffic and predict congestion situations, adjusting the timing of traffic lights and redirecting traffic based on demand. In addition, AI systems can identify and respond to incidents, such as accidents or breakdowns, immediately, facilitating the intervention of emergency services and reducing the impact on traffic (Banister, 2018).

Benefits of Intelligent Traffic Management

The implementation of intelligent traffic management systems offers multiple benefits for both transportation efficiency and the environment. By optimizing vehicle flow, these systems reduce travel times and reduce congestion in urban areas, improving the user experience and making transportation more accessible and predictable. In addition, by reducing the time vehicles spend idling or in constant braking situations, fuel consumption and greenhouse gas emissions are reduced, contributing to urban sustainability goals (Litman, 2021).

Another important benefit of intelligent traffic management is the improvement in road safety . By identifying traffic incidents in real time and adjusting the flow of vehicles, these systems can reduce the risk of accidents and improve safety for all road users. Likewise, prioritizing public transport at traffic lights can make buses and streetcars more efficient and attractive to users, incentivizing the use of sustainable modes of transport and reducing the number of cars on the road (Cohen & Shaheen, 2018).

Challenges in the Implementation of Intelligent Traffic Management Systems

Despite its many benefits, the implementation of intelligent traffic management systems faces several technical, economic and social challenges. One of the main obstacles is the high cost of installing and maintaining the necessary infrastructure, including sensors, cameras, communication systems, and data servers. For many cities, especially those with limited budgets, the initial investment can be a significant impediment (Gossling, 2021). In addition, upgrading older infrastructure to smart systems can be costly and complex.

Another major challenge is data privacy and security. Intelligent traffic management systems collect large amounts of information on travel patterns and vehicle activity, which raises concerns about protecting users' privacy. The implementation of these systems requires advanced security protocols and a regulatory framework that ensures the protection of personal data and minimizes the risk of cyberattacks (Docherty et al., 2018).

Public acceptance also presents a challenge in the adoption of intelligent traffic management technologies. Although the benefits are clear, some people may be reluctant to have their commutes monitored and managed in an automated manner. Transparency in the use of data and effective communication about the benefits of these systems are critical to gaining public trust and ensuring successful adoption.

The Role of Civil Engineers in Intelligent Traffic Management

Civil engineers play an essential role in the design, implementation and maintenance of intelligent traffic management systems. Their work involves not only installing the necessary infrastructure, but also adapting existing urban infrastructure to support new technologies. Civil engineers are responsible for integrating sensors and communication devices into roads, designing traffic systems that respond to dynamic conditions, and ensuring that infrastructure is maintained in optimal condition for safe and efficient operation.

In addition, civil engineers work closely with data analysts and technology specialists to develop algorithms that optimize traffic management and enable real-time decision making. This collaboration is crucial to make the most of the data generated by smart systems and ensure that urban infrastructures respond optimally to the city's needs.

Another important aspect of the work of civil engineers is the planning and design of infrastructure that promotes sustainability. Intelligent traffic management systems not only optimize vehicle flow, but also enable the prioritization of public transport and active mobility, such as cycling and walking, in urban areas. By designing infrastructure that favors these sustainable modes of transport, civil engineers contribute to reducing the carbon footprint of urban transport and creating a more accessible and healthy environment for inhabitants.

Intelligent traffic management is a transformative strategy to improve urban mobility, making transportation more efficient, safe and sustainable. Through advanced technologies, such as artificial intelligence, sensors, and real-time connectivity, these systems make it possible to reduce congestion, lower emissions, and improve road safety. However, their implementation presents technical, economic and social challenges that must be addressed to ensure that the benefits of intelligent traffic management reach all citizens.

For civil engineers, intelligent traffic management is an opportunity to innovate in the design of urban infrastructure that responds to the demands of 21st century mobility . Through careful planning, interdisciplinary collaboration and the use of sustainable technologies, engineers can lead the way in creating cities where traffic is managed in an efficient and environmentally friendly manner, thereby improving the quality of life for residents and moving towards a future of smart urban mobility.

4.5 Intelligent Payment Systems in Urban Transportation: Convenience and Efficiency for Users and Operators

Smart payment systems have revolutionized urban transportation, enabling users to pay quickly, securely and contactlessly. The implementation of advanced digital payment technologies, such as smart cards, mobile apps and contactless payment systems, has not only simplified the user experience, but also optimized the operation of transportation systems, reducing costs and improving efficiency. This approach makes transportation more accessible and convenient, eliminating the need for cash and facilitating the integration of multiple modes of transportation into a single payment platform (Cohen & Shaheen, 2018). This section explores the benefits, challenges, and principles of smart payment systems in urban transportation and the role of civil engineers in implementing this technology.

Principles and Benefits of Smart Payment Systems

Smart payment systems are designed to facilitate access to multiple transportation options through fast and seamless payment methods. One of the key principles of these systems is integration, which allows users to pay for different transportation services-such as buses, trains, bikes, and shared scooters-on a single platform, eliminating the need for separate tickets and improving the user experience (Litman, 2021). Contactless payment technology, such as proximity cards and mobile apps, facilitates the payment process, allowing users to access transportation faster and more conveniently.

The implementation of smart payment systems also benefits transport operators. By reducing the use of cash, operating costs and the risk of loss and theft are reduced. In addition, digital systems enable transportation agencies to collect real-time data on demand and user behavior, which facilitates more efficient service planning and management. For civil engineers, the integration of smart payment systems in urban transportation involves designing infrastructure that can adapt to various payment technologies and improve accessibility and convenience for users (Gossling, 2021).

Integration of Multiple Modes of Transport on a Single Payment Platform

One of the greatest benefits of smart payment systems is their ability to integrate various modes of transportation into a single platform. This integration allows users to plan and pay for their trips seamlessly, combining different transportation options according to their needs, without the need for multiple transactions. This approach falls under the concept of Mobility as a Service (MaaS), which allows users to access multiple modes of transportation from a single application, offering an enhanced user experience and promoting the use of sustainable and shared modes of transportation (Banister, 2018).

The integration of transportation modes into a single payment platform requires a robust infrastructure that supports real-time processing of multiple transactions and ensures the security of user data. In addition, smart payment systems must be accessible to all citizens, regardless of their level of technological literacy or access to digital devices (Smith, 2023). For civil engineers, the challenge is to design payment systems that are inclusive and accessible, ensuring that all users can benefit from a simplified and frictionless payment experience.

Security and Privacy in Digital Payment Systems

Security and privacy are critical aspects in the implementation of smart payment systems. As users conduct transactions via mobile devices or contactless cards, it is critical to ensure that personal and financial information is protected against fraud and unauthorized access. Smart payment systems must be designed with advanced security measures, such as data encryption and multi-factor authentication, that ensure user information is protected at all times (Cohen & Shaheen, 2018).

For civil engineers, this means working in collaboration with cybersecurity experts and software developers to implement solutions that minimize the risk of data breaches and comply with current privacy regulations. In addition, it is important that smart payment systems provide transparency about data usage, allowing users to have control over the information they share and how it is used on the platform.

Accessibility and Equity in Access to Intelligent Payment Systems

Despite their multiple benefits, the implementation of digital payment systems poses the challenge of ensuring that all citizens can access these services in an equitable manner. In many cities, not all users have access to mobile devices or bank cards, which could limit their ability to use smart payment systems. For these systems to be inclusive, it is important to consider alternative payment options, such as prepaid cards and discounts for low-income people (Litman, 2021).

In addition, it is critical to ensure that smart payment systems are accessible in all areas of the city, including those with lower population density or less access to technology. For civil engineers, designing accessible payment systems means developing infrastructure that supports flexible payment methods and is evenly distributed throughout the urban space, ensuring that all citizens can benefit from a convenient and accessible transportation experience.

Smart payment systems are a key tool to improve the user experience in urban transportation, offering a fast, secure and contactless payment alternative that facilitates access to multiple mobility options. By integrating various modes of transportation into a single payment platform, these systems enable smoother and more sustainable mobility, encouraging the use of shared modes of transportation and reducing dependence on the private automobile. For civil engineers, the implementation of smart payment systems is an opportunity to innovate in the design of infrastructures that facilitate an improved user experience and contribute to the operational efficiency of urban transportation.

However, for these systems to succeed, it is critical that they are inclusive, secure and accessible to all citizens. Data protection, accessibility in disadvantaged areas, and equity in access to technology are challenges that must be addressed to ensure that the benefits of smart payment systems reach the entire population. As cities move towards a more digital and connected future, smart payment systems represent a powerful tool to transform urban transportation into an accessible and efficient service, where convenience and sustainability are integrated into a complete mobility experience.

Chapter 5: The City of the Future: Innovation and Sustainability in Urban Planning

The concept of the "city of the future" evokes an urban environment in which technology, sustainability and human well-being are integrated to create spaces that not only respond to today's needs, but also anticipate and adapt their infrastructure and services to the challenges of tomorrow. In a world marked by population growth, climate change and resource scarcity, the urban planning of the future faces the task of designing resilient, inclusive and technologically advanced cities, capable of evolving with their inhabitants and the surrounding environment.

The city of the future is not just a futuristic utopia; it is an achievable goal through the implementation of innovations that are already shaping some of the world's most advanced cities. From smart, self-sufficient buildings to autonomous transportation networks and decentralized renewable energy systems, the elements that make up this vision are emerging as key building blocks for the transformation of urban areas. In this chapter, we will explore how advances in technology and sustainability are changing the principles of urban planning and creating a city model in which urban living is synonymous with balancing human needs and environmental limits.

Technology as a pillar of the city of the future

Technology plays a key role in the development of smarter and more efficient cities. Sensor systems, artificial intelligence, the Internet of Things (IoT) and big data analytics enable cities to collect and analyze information in real time, facilitating decision-making and the optimization of public services. Thanks to these innovations, urban planners can manage traffic more efficiently, improve safety, reduce energy consumption, and respond to emergencies quickly and effectively (Docherty et al., 2018).

In addition, technology is also revolutionizing the construction of intelligent buildings that use sensors and automated systems to regulate temperature, lighting and energy consumption, minimizing their environmental impact. Incorporating these buildings into the urban fabric helps to create cities that consume fewer resources while providing a healthier and more comfortable living environment. For civil engineers, this implies a new paradigm in infrastructure design, in which technology and sustainability combine to create structures that not only serve their inhabitants, but also protect the natural environment (Gossling, 2021).

Sustainability and resilience: cities ready for change

Sustainability is an essential component of the city of the future. Faced with the challenges of climate change, resource depletion and biodiversity loss, cities must adopt practices that reduce their ecological footprint and promote the conservation of surrounding ecosystems. The development of green infrastructure, such as parks, green roofs and rainwater harvesting systems, helps mitigate the effect of heat islands, improves air quality and helps manage precipitation efficiently (Banister, 2018).

Resilience is another fundamental aspect in the design of the city of the future. Urban infrastructures must be prepared to withstand and adapt to extreme events, such as floods, droughts and heat waves, which will become increasingly frequent due to climate change. This requires careful planning and the use of durable and sustainable materials that enable urban infrastructures to withstand the wear and tear of time and adverse weather conditions. For civil engineers, urban resilience represents a technical and logistical challenge, but also an opportunity to develop infrastructures that can protect and serve communities in times of crisis (Rodrigue, 2020).

Inclusion and quality of life in planning for the future

The city of the future will not only focus on technology and sustainability, but will also put people at the center of its planning. Inclusion and quality of life are essential aspects in the design of modern cities that seek to improve the well-being of all their inhabitants, regardless of their socioeconomic status or location. The creation of accessible public spaces, the promotion of active mobility and access to quality basic services are fundamental elements in the construction of fairer and more equitable cities.

Future urban planning must ensure that all citizens have access to development opportunities and a healthy living environment. This implies the creation of policies that promote equity in access to services such as education, health, transportation and housing. For civil engineers, the design of inclusive and accessible infrastructure is a priority in the development of cities of the future, as this infrastructure not only improves the quality of life of the inhabitants, but also fosters social cohesion and a sense of community.

Innovation and the role of civil engineers in the city of the future

The role of civil engineers in building the city of the future is fundamental. Their work involves not only the construction of infrastructure, but also the integration of sustainable technologies and the planning of systems that respond to the demands of an ever-changing society. Civil engineering must adapt to a new paradigm in which infrastructures not only fulfill a technical function, but also promote sustainability, resilience and the well-being of citizens (Litman, 2021).

In addition, civil engineers must lead in the implementation of innovations such as near-zero energy buildings, autonomous transportation systems, distributed energy networks, and advanced waste management infrastructures. The city of the future requires a multidisciplinary approach and close collaboration between engineers, urban planners, architects, and technology and environmental experts. For civil engineers, this means not only adopting new technologies, but also fostering a holistic view of urban planning that considers both human needs and the limits of the planet.

Building the city of the future is not a short-term project; it is a commitment to the well-being of present and future generations. For civil engineers, this vision represents a unique opportunity to contribute to the development of cities that not only respond to today's challenges, but also anticipate and adapt to the needs of an ever-changing world. With a focus on innovation, sustainability and inclusion, the city of the future is more than an ideal; it is the next step towards an urban environment that serves people and the planet.

5.1 Smart Infrastructures: Adaptation and Efficiency in the City of the Future.

Smart infrastructures are at the heart of the city of the future. These structures not only fulfill traditional functions, such as transportation, resource management and the provision of basic services, but also incorporate advanced technologies to optimize their performance, adapt to the changing needs of cities and reduce their environmental impact. Equipped with sensors, monitoring systems, and automated response capabilities, smart infrastructures offer innovative solutions to address the challenges of urbanization, climate change, and sustainability (Docherty et al., 2018). This section discusses how these infrastructures are transforming urban planning, their benefits, challenges, and the critical role of civil engineers in their development.

Characteristics and Benefits of Intelligent Infrastructures

Smart infrastructures are characterized by their ability to collect, analyze and act on data in real time. Through the use of sensors, the Internet of Things (IoT) and artificial intelligence (AI) systems, these infrastructures can monitor their status and performance, detect failures and optimize their functions autonomously. For example, a smart bridge equipped with sensors can alert about structural problems before they pose a risk to users, allowing preventive maintenance and reducing costs associated with major repairs (Rodrigue, 2020).

One of the main benefits of smart infrastructures is their ability to improve resource efficiency. Systems such as smart grids optimize electricity distribution and consumption, reducing losses and promoting the use of renewable energies. Likewise, smart infrastructures can improve water management through leak detection and efficient distribution systems, ensuring sustainable access to the resource (Gossling, 2021). For civil engineers, the design and implementation of these infrastructures represent an opportunity to lead in the creation of more efficient, sustainable and resilient cities.

Transportation and Urban Mobility Applications

In transportation, smart infrastructures are revolutionizing urban mobility. Systems such as smart traffic lights, charging stations for electric vehicles, and dedicated lanes for autonomous vehicles optimize traffic flow and improve the user experience. For example, smart traffic lights, connected to traffic monitoring networks, adjust their cycles in real time to minimize congestion and reduce waiting times (Litman, 2021). This type of technology not only improves transportation efficiency, but also reduces greenhouse gas emissions associated with traffic jams.

Smart charging stations for electric vehicles represent another key application of smart infrastructure. These stations are equipped with technologies that enable fast and efficient charging, and can be integrated with renewable energy grids to minimize their carbon footprint. In addition, smart transportation infrastructures enable the collection of data on mobility patterns, which can be used to plan more efficient and accessible public transportation routes (Cohen & Shaheen, 2018).

Challenges in the Development of Intelligent Infrastructures

Despite their numerous benefits, smart infrastructures face several challenges in their development and implementation. One of the biggest hurdles is the initial installation cost, which can be significantly higher than that of conventional infrastructures. This cost includes not only materials and construction, but also advanced technological systems and the training of personnel needed to operate them (Banister, 2018). For many cities, especially those with limited resources, this factor can be a major impediment.

Another key challenge is data security. Smart infrastructures collect large amounts of information about their users and their environment, which poses risks in terms of privacy and cybersecurity. Systems must be designed with advanced protection measures to ensure that data is secure and infrastructures are protected against potential cyber attacks (Docherty et al., 2018).

In addition, interoperability between different systems and technologies is another critical challenge. Smart infrastructures require seamless integration between multiple systems, such as transportation networks, utilities and digital platforms, which can be complicated by the diversity of standards and technologies used by different vendors. For civil engineers, this means working closely with technology experts and regulators to ensure that infrastructures are compatible and operate cohesively.

Role of Civil Engineers in Intelligent Infrastructures

Civil engineers are central to the design, development and implementation of smart infrastructure. Their work ranges from the planning and construction of physical structures to the integration of advanced technologies that enable these infrastructures to adapt and respond to urban needs. This includes incorporating sensors, monitoring systems and communication networks into roads, bridges, buildings and other critical infrastructure.

In addition, civil engineers must ensure that these infrastructures are sustainable and resilient. This involves using durable materials with low environmental impact, as well as designing systems that can withstand extreme weather conditions and adapt to future changes in urban needs. The ability of civil engineers to innovate and lead in the development of smart infrastructure will be essential to the success of the city of the future (Rodrigue, 2020).

Smart infrastructures are an essential component of the city of the future, offering innovative solutions to the challenges of urbanization, climate change and sustainability. Through advanced technologies such as IoT, artificial intelligence and real-time monitoring systems, these infrastructures improve efficiency, reduce costs and promote a more sustainable use of resources. However, their development also poses technical, economic and security challenges that must be carefully addressed.

For civil engineers, the design and implementation of smart infrastructure represents an opportunity to lead in creating more resilient, sustainable and livable cities. As cities evolve into a more technological and connected future, smart infrastructure will not only transform the way we live and work, but also lay the foundation for an urban environment that proactively responds to the needs of its inhabitants and the planet.

5.2 Sustainable and Smart Buildings: Pillars of the City of the Future

Sustainable and intelligent buildings are key elements in the planning and development of the city of the future. These buildings not only seek to reduce their environmental impact through the efficient use of resources, but also integrate advanced technology to optimize their operation, improve the quality of life of their occupants, and adapt to the changing needs of the urban environment. Equipped with automated management systems, sensors, and advanced technologies, sustainable and smart buildings represent a new paradigm in architecture and engineering, where efficiency, sustainability, and connectivity converge to redefine the built environment (Docherty et al., 2018). This section explores the benefits, characteristics, and challenges of these buildings, as well as the role of civil engineers in their design and implementation.

Characteristics of Sustainable and Smart Buildings

Sustainable buildings are designed to minimize their environmental impact throughout their life cycle, from construction to operation and demolition. This includes the use of recycled and low environmental impact materials, the incorporation of energy-efficient systems, and the implementation of water management technologies that reduce consumption and maximize reuse (Gossling, 2021). For example, rainwater harvesting and recycling systems allow buildings to reduce their dependence on external sources, while solar panels and heat pumps contribute to meeting their energy needs through renewable sources.

On the other hand, smart buildings incorporate advanced technology, such as IoT sensors and artificial intelligence systems, to monitor and manage the consumption of energy, water and other resources in real time. These systems automatically optimize internal conditions, such as lighting and climate control, to ensure occupant comfort while reducing energy consumption. In addition, smart buildings are designed to integrate with urban infrastructure, enabling communication with smart grids and urban transportation systems to improve overall city efficiency (Litman, 2021).

Benefits of Sustainable and Smart Buildings

Sustainable and intelligent buildings offer multiple environmental, economic and social benefits. From an environmental perspective, these buildings help reduce greenhouse gas emissions and the carbon footprint associated with urban activities. By using energy-efficient technologies and renewable sources, smart buildings minimize energy consumption and contribute to climate change mitigation (Cohen & Shaheen, 2018). In addition, by integrating water management systems, these buildings promote the conservation of this vital resource, which is especially relevant in water-stressed regions.

In economic terms, sustainable and intelligent buildings generate significant savings in operating costs due to their energy efficiency and efficient use of resources. Although the initial investment may be higher, the long-term benefits, such as lower energy bills and reduced maintenance, far outweigh these costs (Rodrigue, 2020). Also, these buildings increase property values and attract tenants and buyers looking for modern, efficient and environmentally friendly spaces.

From a social perspective, intelligent buildings improve the quality of life of their occupants by providing healthy, comfortable and customized spaces. For example, advanced ventilation systems improve indoor air quality, while automated lighting and climate control create a more pleasant and productive environment. In addition, these buildings promote inclusion and accessibility by incorporating universal designs that ensure that everyone, regardless of physical ability, can enjoy their facilities.

Challenges in the Development of Smart and Sustainable Buildings

The development of sustainable and intelligent buildings faces several technical, economic and regulatory challenges. One of the main obstacles is the high initial cost associated with advanced technologies and sustainable materials. While these costs are often amortized over time, they can be prohibitive for developers with limited resources. To address this challenge, it is critical that governments and financial institutions provide incentives and financing programs that encourage the construction of sustainable buildings (Banister, 2018).

Another challenge is the technical complexity associated with integrating advanced technologies into the design and operation of buildings. Smart systems require a robust digital infrastructure, as well as trained personnel for installation and maintenance. In addition, interoperability between different systems and platforms can be an issue, especially in large, complex buildings. For civil engineers, this means working closely with technology experts and software developers to ensure that systems are compatible and operate efficiently.

Regulation is also a challenge, as building regulations and urban codes may not be updated to address the specific requirements of sustainable and smart buildings. This can delay the approval and construction of these buildings, especially in regions where innovation in the building sector is not yet fully supported by public policy. To overcome this obstacle, it is necessary for civil engineers and developers to work collaboratively with policymakers to update regulations and promote standards that support sustainability and innovation.

Role of Civil Engineers in the Design of Sustainable and Smart Buildings

Civil engineers play an essential role in the design and construction of sustainable and intelligent buildings. Their work ranges from selecting sustainable materials to designing structural and service systems that minimize resource consumption and maximize efficiency. In addition, civil engineers are responsible for integrating advanced technologies into building design, ensuring that smart systems function harmoniously and contribute to the overall goal of sustainability (Rodrigue, 2020).

Another crucial aspect of the role of civil engineers is to ensure that buildings are resilient in the face of challenges such as climate change and extreme events. This involves designing structures that can withstand events such as floods, earthquakes and hurricanes, while ensuring the safety and well-being of their occupants. The ability of civil engineers to innovate and lead in the construction of sustainable and intelligent buildings is critical to the success of the city of the future.

Sustainable and intelligent buildings are essential components of the city of the future, offering innovative solutions to the environmental, economic and social challenges of urban areas. Through advanced technologies and sustainable design practices, these buildings not only reduce their environmental impact, but also improve the quality of life of their occupants and generate significant economic benefits. However, their development requires overcoming technical, economic and regulatory challenges that demand collaboration between civil engineers, urban planners, developers and policy makers.

For civil engineers, sustainable and intelligent buildings represent a unique opportunity to contribute to the development of more resilient, efficient and inclusive cities. With a focus on sustainability, innovation and human well-being, the buildings of the future will not only be physical structures, but also powerful tools for the transformation of our cities and the creation of an urban environment that serves people and the planet.

5.3 Renewable Energy and Smart Grids: Powering the City of the Future

Renewable energy and smart grids play a central role in the development of the city of the future. As cities face increasing energy demands and the effects of climate change, the transition to clean energy sources and the implementation of smart grids have become critical priorities. This approach not only reduces greenhouse gas emissions, but also improves energy efficiency, encourages decentralization of energy generation, and increases the resilience of cities to extreme events (International Energy Agency [IEA], 2022). This section explores the advantages, challenges and applications of renewable energy and smart grids in the urban context, as well as the role of civil engineers in their implementation.

Transition to Renewable Energy in Cities

The adoption of renewable energy sources, such as solar, wind and geothermal, is essential to reduce dependence on fossil fuels and minimize the environmental impact of cities. Solar panels and wind turbines, for example, are being integrated into buildings and urban spaces to generate clean, decentralized energy. This renewable energy not only decreases carbon emissions, but also allows cities to tap into local resources, reducing the need to import energy from external sources (Litman, 2021).

In addition, renewables are key to fostering economic sustainability. Although the initial investment can be high, operating costs are significantly lower than those of fossil fuels, and the availability of government incentives and tax breaks makes renewable solutions increasingly accessible. For civil engineers, this means designing urban infrastructure that integrates renewable energy sources, from solar roofs on buildings to urban wind farms, and ensuring that these solutions are safe and efficient.

Smart Grids: The Connection between Energy and Technology

Smart grids are advanced electricity distribution systems that use digital technologies to manage energy supply and demand in real time. These grids not only improve energy efficiency by minimizing transmission losses, but also enable more effective integration of renewable energy sources into the power system. For example, a smart grid can balance solar or wind power generation with the needs of users, ensuring a constant supply even in variable weather conditions (Cohen & Shaheen, 2018).

Another key benefit of smart grids is their ability to increase the resilience of cities to power outages and extreme events. By collecting and analyzing real-time data, these grids can identify problems before they become major failures, reroute power as needed, and restore supply quickly in the event of outages. For civil engineers, planning and implementing smart grids requires a multidisciplinary approach, combining expertise in electrical infrastructure, information technology and urban sustainability.

Renewable Energy and Smart Grid Integration

The integration of renewable energy and smart grids is essential to maximize the benefits of both systems. Smart grids allow cities to more efficiently manage energy generated by renewable sources, storing excess energy in batteries and distributing it when demand is high. In addition, these grids facilitate the active participation of citizens in energy management, allowing them to monitor their consumption and contribute to the grid through distributed generation, such as residential solar power (Banister, 2018).

The electrification of transportation also benefits from this integration. Electric vehicle charging systems can be connected to smart grids to optimize energy consumption, charging vehicles during off-peak hours and using renewable sources wherever possible. For civil engineers, this means designing charging stations and distribution networks that are flexible and able to adapt to fluctuations in energy generation and demand.

Renewable Energy and Smart Grid Implementation Challenges

Despite their numerous benefits, the implementation of renewable energy and smart grids faces significant challenges. One of the main obstacles is the initial investment, which can be high for both renewable generation infrastructure and grid modernization. In addition, the integration of multiple energy sources into a unified grid requires advanced management and coordination systems, which adds technical complexity to the process (Rodrigue, 2020).

Another challenge is to ensure equity in access to these technologies. In many cities, low-income communities face economic and technological barriers that make it difficult for them to participate in the energy transition. To address this problem, it is critical that public policies promote incentives and support programs that enable all communities to benefit from renewable energy and smart grids.

Finally, data security and privacy represent a major concern. Smart grids collect vast amounts of information about users, posing cybersecurity and data protection risks. For civil engineers, this means working collaboratively with technology experts and regulators to ensure that systems are secure and comply with privacy regulations.

The Role of Civil Engineers in the Energy Transition

Civil engineers play a crucial role in the implementation of renewable energy and smart grids in cities. Their work includes the design and construction of energy generation and storage infrastructure, such as solar panels, wind turbines and charging stations for electric vehicles. In addition, civil engineers are responsible for integrating these solutions into the urban fabric so that they are accessible, safe and aesthetically appropriate.

Another important aspect is the planning and development of smart grids that can efficiently manage the generation and distribution of renewable energy. This requires close collaboration with technology experts, grid operators and local authorities to ensure that the solutions implemented are sustainable and scalable.

Renewable energy and smart grids are fundamental pillars of the city of the future, offering innovative solutions to the energy and environmental challenges of urban areas. By promoting sustainability, efficiency and resilience, these technologies transform the way cities generate and consume energy, creating an environment that is cleaner, more equitable and adapted to the needs of the 21st century.

For civil engineers, the transition to renewable energy and smart grids represents a unique opportunity to lead in the creation of sustainable and resilient cities. With a focus on innovation, equity and collaboration, engineers can help build a future in which cities are not only functional and technological, but also environmentally responsible and focused on the well-being of their inhabitants.

5.4 Waste Management and Circular Economy: A New Vision for Cities of the Future

Efficient waste management is a crucial challenge for modern cities, especially in a world where population growth and excessive consumption have led to an exponential increase in waste generation. The city of the future seeks not only to minimize waste, but also to transform it into resources through the implementation of strategies based on the circular economy. This approach promotes the design of urban systems that reduce, reuse and recycle materials, closing the life cycle of products and reducing dependence on virgin raw materials (Ellen MacArthur Foundation, 2020). This section analyzes the strategies, benefits and challenges of waste management in the city of the future, as well as the fundamental role of civil engineers in its implementation.

Waste Management Strategies in the City of the Future

Waste management in the cities of the future focuses on reducing waste at source and creating infrastructure for its efficient treatment. One of the key strategies is the development of smart collection and sorting systems, which use advanced technologies such as IoT sensors and data analytics to optimize collection routes, monitor waste volume, and ensure proper segregation of materials (Docherty et al., 2018).

Another strategy is the promotion of advanced recycling plants and composting centers, which allow organic waste and recyclable materials to be treated efficiently. These facilities not only help reduce the amount of waste sent to landfills, but also generate valuable products, such as fertilizers and secondary raw materials. In addition, the energetic use of waste, through technologies such as incineration with energy recovery and anaerobic digestion, allows waste to be converted into electricity, heat and biogas, offering a sustainable solution for the management of non-recyclable waste (International Solid Waste Association [ISWA], 2021).

The Circular Economy as an Urban Model

The circular economy is an approach that seeks to transform the linear system of "produce, consume and dispose" into a model where resources are kept in use for as long as possible. This includes the design of durable and repairable products, the reuse of materials and the creation of efficient recycling systems that allow resources to be reintroduced into the production chain (Ellen MacArthur Foundation, 2020).

In the urban context, the circular economy encourages the integration of sustainable solutions in city planning. This includes designing buildings with recyclable materials, reusing water in closed systems, and creating urban infrastructures that facilitate the recycling and recovery of materials. For civil engineers, the circular economy implies a change in the approach to design and construction, prioritizing sustainable materials and innovative solutions that maximize the use of resources.

Benefits of Waste Management and the Circular Economy

The adoption of advanced waste management and circular economy strategies offers multiple environmental, economic and social benefits. From an environmental perspective, these practices reduce pressure on landfills, reduce soil and water pollution, and reduce greenhouse gas emissions associated with waste decomposition and natural resource extraction (Litman, 2021).

Economically, the recovery of materials and the generation of energy from waste create opportunities for the development of new industries and green jobs. In addition, the implementation of the circular economy fosters innovation in product and process design, increasing the competitiveness of local businesses. Socially, these strategies improve the quality of life by reducing the problems associated with waste accumulation, such as odors, pests and public health risks.

Challenges in the Implementation of Circular Waste Management Systems

Despite its benefits, the transition to circular waste management systems faces several challenges. One of the main obstacles is the lack of adequate infrastructure for waste collection, sorting and treatment, especially in cities with limited resources. In addition, the implementation of advanced technologies requires significant investments, which can be an impediment for many urban communities (Rodrigue, 2020).

Another important challenge is the need to change citizens' habits and behaviors.

The adoption of the circular economy depends to a large extent on the collaboration of residents, who must actively participate in waste segregation and consumption reduction. To overcome this challenge, it is essential to implement education and awareness campaigns that promote sustainable practices and foster a culture of environmental responsibility.

The integration of public policies is also essential to support the circular economy. This includes creating regulations that incentivize reuse and recycling, as well as setting clear targets for waste reduction and materials recovery. For civil engineers, this involves collaborating with governments and other stakeholders to ensure that proposed infrastructure and solutions are viable and sustainable.

The Role of Civil Engineers in Waste Management and the Circular Economy

Civil engineers play a crucial role in the implementation of advanced waste management systems and circular economy in the cities of the future. Their work includes the design of waste treatment infrastructures, such as recycling plants, composting centers and energy recovery systems. In addition, civil engineers are responsible for developing efficient waste collection and transportation networks that minimize emissions and maximize operational efficiency.

Another key aspect is the design of urban infrastructure that incorporates circular economy principles. This includes the selection of recyclable materials in the construction of buildings and roads, as well as the design of water and energy reuse systems. The ability of civil engineers to innovate and lead in the development of sustainable solutions is essential to the success of waste management and the circular economy.

Waste management and the circular economy are key elements in building sustainable, resilient and efficient cities. Through the implementation of advanced technologies and comprehensive strategies, the cities of the future can transform waste into resources, reduce their environmental impact and generate significant economic and social benefits. However, achieving this transition requires overcoming technical, economic and cultural challenges that demand the collaboration of all stakeholders.

For civil engineers, the circular economy represents a unique opportunity to lead in the design and construction of urban systems that optimize the use of resources and minimize waste. With a focus on innovation, sustainability and collaboration, civil engineers can help build a future in which cities are not only functional, but also environmentally responsible and focused on the well-being of their inhabitants.

References

Inter-American Development Bank (IDB). (2023). *Transforming cities: Transportation-oriented development.* Inter-American Development Bank. https://blogs.iadb.org/ciudades-sostenibles/es/transformando-ciudades-desarrollo-transport-oriented/

Banister, D. (2018). *Transport, climate change and the city.* Routledge. https://www.taylorfrancis.com/books/mono/10.4324/9780203074435/transport-climate-change-city-david-banister-robin-hickman.

Cohen, A., & Shaheen, S. (2018). *Planning for shared mobility.* American Planning Association. https://www.planning.org/publications/report/9107556/

Docherty, I., Marsden, G., & Anable, J. (2018). The governance of smart mobility. *Transportation Research Part A: Policy and Practice, 115,* 114-.
125.https://www.sciencedirect.com/science/article/abs/pii/S0965856417312908

Ellen MacArthur Foundation (2020). *Circular economy: A systems solution framework.* https://ellenmacarthurfoundation.org/topics/circular-economy-introduction/overview

Fernández, P. (2023). *Transport systems and urban sustainability*. Editorial Técnica. https://enlacealafuente.com

García, A. (2023). *Transportation systems and urban mobility*. Editorial Urbana. https://enlacealafuente.com

Garcia, J. (2023). *Design of sustainable transport systems*. Editorial Sostenibilidad Urbana. https://enlacealafuente.com

Gehl, J. (2010). *Cities for people*. Island Press.
https://islandpress.org/books/cities-people

Gómez, L. (2023). *Sustainable transport: Challenges and opportunities for the cities of the future.* Editorial Verde. https://enlacealafuente.com

Gossling, S. (2021). *Urban transport justice.* Routledge.

https://www.taylorfrancis.com/books/mono/10.4324/9780429287558/urban-transport-justice-stefan-g%C3%B6ssling

Hernández, L. (2023). *Mobility and urbanism: Towards sustainable and equitable cities*. Editorial Innovación Urbana. https://enlacealafuente.com

International Energy Agency (IEA). (2022). *The role of sustainable transport in reducing urban pollution* https://www.iea.org/reports/sustainable-transport

International Solid Waste Association (ISWA) (2021). *Global waste management outlook.* https://www.iswa.org/programmes/knowledge-base/gwmo/

Jacobs, J. (1961). *The death and life of great American cities*. Random House. https://www.penguinrandomhouse.com/books/611123/the-death-and-life-of-great-american-cities-by-jane-jacobs/

Litman, T. (2021). *Introduction to transportation demand management: Planning and evaluating mobility management solutions*. Victoria Transport Policy Institute. https://www.vtpi.org/tdm/tdmintro.pdf

Litman, T. (2021). *Planningfor sustainable transportation: Indicators and strategies.* Victoria Transport Policy Institute.
https://www.vtpi.org/plansus.pdf

López, R. (2022). *Mobility and urban development: The impact of transportation systems* . Editorial Movilidad. https://enlacealafuente.com

Martínez, J. (2021). *Transportation engineering and urban mobility*. Editorial Técnica. https://enlacealafuente.com

Moreno, C., Allam, Z., Chabaud, D., Gall, C., & Pratlong, F. (2021). Introducing the "15-minute city": Sustainability, resilience and place identity in future post-pandemic cities. *Smart Cities, 4*(1), 93-111https://www.mdpi.com/2624-6511/4/1/6.

Newman, P., & Kenworthy, J. (2015). *The end of automobile dependence: How cities are moving beyond car-based planning*. Island Press.
https://islandpress.org/books/end-automobile-dependence

UN-Habitat (2023). *Planning and design for sustainable urban mobility.* UN-Habitat. https://unhabitat.org/planificacion-y-diseno-de-una-movilidad-urbana-sustainable-en-espanol-language-version

World Health Organization (WHO). (2021). *Air pollution and health.* https://www.who.int/health-topics/air-pollution#tab=tab 1

Pérez, M. (2020). *Transport and accessibility: Towards more inclusive cities.* Editorial Movilidad Inclusiva. https://enlacealafuente.com

Pucher, J., & Buehler, R. (2019). *City cycling.* MIT Press.
https://mitpress.mit.edu/9780262527587/city-cycling/

Ramírez, G. (2022). *Urban mobility: Transportation systems and sustainability.* Editorial Urbana. https://enlacealafuente.com

Rodrigue, J. P. (2020). *The geography of transport systems* (5th ed.). Routledge. https://transportgeography.org/?page id=1121

Rodriguez, S. (2024). *The future of urban mobility: AI, autonomous vehicles and MaaS.* Editorial Innovación. https://enlacealafuente.com

Smith, J. (2023). *Smart payment systems in urban transportation: Challenges and solutions.*
Urban Mobility Press. https://www.urbanmobilitypress.com/smart-payment-systems

World Economic Forum (2022). *The future of urban mobility: Towards smarter and more sustainable cities.* World Economic Forum. https://www.weforum.org/reports/the-future-of-urban-mobility.

Printed by Books on Demand GmbH, Norderstedt / Germany